Linear and Switch-Mode RF Power Amplifiers

Linear and Switch-Mode RF Power Amplifiers

Design and Implementation Methods

Abdullah Eroglu

CRC Press
Taylor & Francis Group
Boca Raton London New York

CRC Press is an imprint of the
Taylor & Francis Group, an **informa** business

CRC Press
Taylor & Francis Group
6000 Broken Sound Parkway NW, Suite 300
Boca Raton, FL 33487-2742

© 2018 by Taylor & Francis Group, LLC
CRC Press is an imprint of Taylor & Francis Group, an Informa business

No claim to original U.S. Government works

Printed on acid-free paper

International Standard Book Number-13: 978-1-4987-4576-5 (Hardback)
978-1-1387-4577-3 (Paperback)

Visit the Taylor & Francis Web site at
http://www.taylorandfrancis.com

and the CRC Press Web site at
http://www.crcpress.com

*Dedicated to my family who has always shown
great patience and support for my research.*

Contents

Preface

Radio frequency power amplifiers are critical part of communication, semiconductor wafer processing, magnetic resonance imaging (MRI), and radar systems. The amplifier system not only requires expertise in the design of amplifier topologies, but it also requires knowledge in the design of its surrounding components. The authors' previous two books, *Introduction to RF Circuit Design Techniques for MF-UHF Applications* and *Introduction to RF Power Amplifier Design and Simulation*, have presented the design and simulation of the amplifier components and devices. This book is written to provide the implementation methods for RF amplifiers, which can operate in linear or switch mode via several real-world engineering problems. Step-by-step design methods to design amplifiers are exemplified with several design problems. MATLAB, Matchcad Pspice, and ADS simulation tools are used to aid the design process. All the design examples given include analytical design, simulation verification, and measurement results of the built prototype.

The scope of each chapter in this book can be summarized as follows. Chapter 1 gives introduction to RF Power amplifiers and topologies and discussion on passive and active components and surrounding devices for RF power amplifiers. Chapter 2 discusses two-port parameters including *Z*, *Y*, *h*, *ABCD*, and scattering parameters, which can be used to design small-signal amplifiers. In Chapter 3, impedance matching and resonant networks for amplifiers are detailed. In Chapter 4, small-signal amplifier design methods are introduced with simulation and implementation examples. Chapter 5 discusses design, analysis, and implementation of large-signal amplifier design methods. In this chapter, RF linear amplifiers and different biasing schemes have been presented. The design and implementation of Classes A, B, and AB are given with examples. Furthermore, Class C amplifier design in different modes is also detailed.

In Chapter 6, the design and implementation of RF switch-mode amplifiers including Classes D, E, DE, F, and S topologies have been discussed with application and implementation examples. Generic MATLAB and Mathcad design algorithms to design Classes D, DE, E, and F amplifiers are given. Chapter 7 presents phase-controlled switch-mode amplifiers using harmonic modeling. RF pulsing amplifier design techniques and implementation methods are also discussed in this chapter. Characterization methods for amplifiers have been illustrated using load-pull technique. In Chapter 8, distortion and modulation effects for amplifiers and the method for elimination of the impairments causing these defects have been detailed.

It is important to note that engineering application examples are given in every chapter for each subject to solidify the theory and help designers have hands-on experience of the design problems. Implementation of RF amplifiers is given step by step to make it possible for professionals and students use this book as the main resource.

The MathWorks, Inc.
3 Apple Hill Drive
Natick, MA 01760-2098 USA
Tel: 508-647-7000
Fax: 508-647-7001
E-mail: info@mathworks.com
Web: www.mathworks.com

Acknowledgments

I would like to thank my wife, G. Dilek, for her greatest support, including editorial corrections in the book.

The completion of this book was possible during my sabbatical leave from my home institution, Purdue University Fort Wayne, and I thank it for the support. I would also like to thank University of Gavle, Sweden for hosting me during some portion of that period, which helped me complete this project.

I would also like to thank my editor, Nora Konopka, for her patience and support during the course of the preparation and publication of this book.

Author

Abdullah Eroglu is a professor of electrical engineering at the Electrical and Computer Engineering Department of Purdue University, Fort Wayne. He previously worked as a radio frequency (RF) senior design engineer at MKS Instruments, ENI Products, Rochester, New York, and as a faculty fellow in the Fusion Energy Division of Oak Ridge National Laboratory, Oak Ridge, Tennessee.

His research focuses on design and development of RF devices for applications including communication, energy, semiconductor manufacturing, and health care. He also investigates the propagation and radiation characteristics of materials for microwave and RF applications.

He earned a PhD in electrical engineering at Syracuse University, Syracuse, New York. He is a recipient of several research awards, including 2013 IPFW Outstanding Researcher Award, 2012 IPFW Featured Faculty Award, 2011 Sigma Xi Researcher of the Year Award; and 2010 IPFW College of Engineering, Technology, and Computer Science Excellence in Research Award.

Dr. Eroglu is the author of five books and has published more than 100 peer-reviewed journals and conference papers. He has several patents in RF component, device design, and methodology. He also serves as a reviewer and editorial board member for several journals.

1 Radio Frequency Amplifier Basics

1.1 INTRODUCTION

Radio frequency (RF) power amplifiers are one of the critical elements for communication systems when signal in use needs to be amplified for several applications. The simplified RF power amplifier block diagram is shown in Figure 1.1.

Radio frequency power amplifiers consist of several surrounding components. The main component of the amplifier is the active device that is critical in the amplification process. The surrounding components in the amplifier are matching networks that are implemented at the input and output of the amplifier, biasing network, stability networks, filters that are used to eliminate spurious or harmonic contents, and control system of the amplifier to ensure the functionality of the amplifier against various conditions.

The type and class of the amplifier are important parameters in the design process, apart from the design of its surrounding components. The amplifier type can be small signal or large signal based on the design, whereas class of the amplifier might belong to a group of linear or switch-mode class of amplifiers. The type and class of the amplifier are based on the application and requirements identified beforehand for the designer. In this book, different components, types, and classes of the amplifiers; mathematical and computer tools that are used to design, simulate, and implement RF; and microwave amplifier and components will be discussed.

1.2 TYPES OF AMPLIFIERS: SMALL SIGNAL AND LARGE SIGNAL AMPLIFIERS

Amplifiers have inherently nonlinear characteristics owing to active devices that are used during amplification process. However, when signal levels are small enough, it does not cause distortion to vary parameters of the active devices, which results in the linear relationship between input and output of the amplifier.

When the conditions satisfy the requirements of the linear approximation, amplifiers can be analyzed using two-port parameters and network methods. The type of the amplifiers in this group are called as small signal amplifiers. The efficiency of small signal amplifiers is not critical. The design for this type of amplifiers use complex conjugate matching method.

For high-power applications, large signal levels are applied for amplification, and this causes distortion in the amplifier due to variation of the active device parameters. Hence, small signal analysis and techniques cannot be applied anymore for these types of amplifiers. The type of the amplifiers in this group are called as large signal amplifiers. The conventional conjugate matching method also does not deliver

1

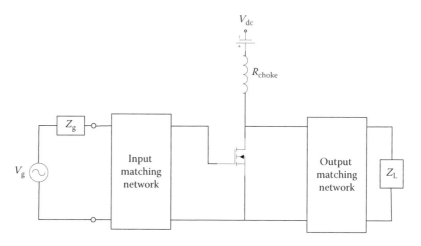

FIGURE 1.1   Basic block diagram of RF power amplifiers.

the maximum power to the load connected at the output of the amplifier. Hence, the technique called load line matching [1] needs to be implemented to overcome the limitations due to physical rating of active device and have the maximum voltage and current swing. The optimum load impedance that will give the practical maximum power can be obtained from the maximum swing points for voltage and current from

$$R_{opt} = \frac{V_{max\_swing}}{I_{max\_swing}} \tag{1.1}$$

The difference between conjugate matching used in small signal amplifier design and load matching used in large signal amplifier design can be better understood using the simplified circuit given in Figure 1.2.

The maximum power transfer in the circuit shown in Figure 1.2 occurs when

$$R_g = R_L \text{ and } B_g = -B_l. \tag{1.2}$$

If we assume that there are no device or component limitations in the circuit, then this will hold true, and the maximum power will be transferred as indicated for small

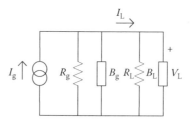

FIGURE 1.2   Simplified circuit for large signal and small signal analysis.

signal analysis. However, if there is limitation for voltage or current, this cannot be applied. For instance, for the circuit in Figure 1.2, let us assume $R_g = 50\ [\Omega]$ and the maximum operating voltage and current are $V_{max} = 20\ [V]$, $I_{max} = 2\ [A]$

Then, the maximum power transfer using conjugate method dictates that

$$R_L = R_L = 50\ [\Omega]$$

Since, R_1 and R_g are in parallel, the equivalent impedance is $R_{eq} = 25\ [\Omega]$. The output voltage is then calculated to be

$$V_L = I_{max} R_{eq} = (2)(25) = 50\ [V]\ >\ V_{max},$$

which is more than the amount of maximum voltage that can be handled in the circuit. Hence, the maximum current that can be obtained with this configuration by staying within the given operating limit of the voltage is $I_{L,max} = 0.8\ [A]$. Then, the maximum power without exceeding the operating limit is

$$P_{L,max} = I_{L,max} V_{L,max} = (0.8)(20) = 16\ [W].$$

Now, let us assume, we would like to use the load line matching method, which is the viable design method for large signal amplifiers. Then, the optimum load resistance is calculated to be

$$R_L = \frac{V_{max}}{I_{max}} = \frac{20}{2} = 10\ [\Omega]$$

Then, the power that can be obtained with this configuration is

$$P_{L,max} = I_{L,max} V_{L,max} = (2)(20) = 40\ [W].$$

This is the method that can now utilize the ratings of devices and gives maximum power attainable in practice.

Large signal amplifiers can be classified as Class A, Class B, and Class AB for linear mode of operation and Class C, Class D, Class E, and Class F for nonlinear mode of operation. Classes D, E, and F are also known as switch-mode amplifiers.

1.3 LINEAR AMPLIFIERS

Consider the amplifier voltage transfer characteristics illustrated in Figure 1.3. The output of the amplifier is linearly proportional to the input of the amplifier via voltage gain β. The output voltage will be identical to the input voltage. The relationship between input and output of the amplifier can be expressed mathematically as

$$v_o(t) = \beta v_i(t). \tag{1.3}$$

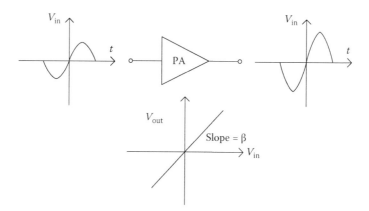

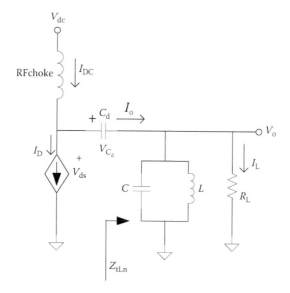

FIGURE 1.3 Linear amplifier voltage transfer characteristics.

FIGURE 1.4 Equivalent circuit representation of linear amplifier mode of operation.

When the transistor is operated as a voltage-dependent current source, linear operational mode for the amplifiers can be obtained. The conduction angle, θ, is then used to determine the class of the amplifier [2]. The conduction angle varies up to 2π based on the amplifier class. The use of transistor as a dependent current source represents linear mode of operation, which is shown in Figure 1.4.

The conduction angles, bias, and quiescent points for linear amplifier are illustrated in Table 1.1. Conduction angle, θ, is defined as the duration of the period in which the given transistor is conducting. The full cycle of conduction is considered to be 360°. The points of intersection with the load line are known as the "quiescent"

TABLE 1.1

Conduction Angles, Bias, and Quiescent Points for Linear Amplifiers

Class	Bias Point	Quiescent Point	Conduction Angle
A	0.5	0.5	2π
AB	0–0.5	0–0.5	π–2π
B	0	0	π
C	<0	0	0–π

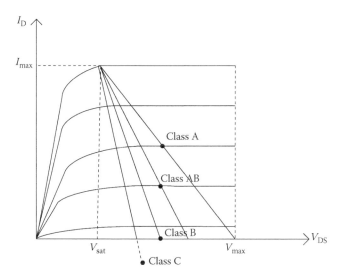

FIGURE 1.5 Load lines and bias points for linear amplifiers.

conditions or "Q points" or the dc bias conditions for the transistor, which represent the operational device voltages and drain current as shown in Figure 1.5.

Classes A, AB, and B amplifiers have been used for linear applications where amplitude modulation (AM), single-sideband modulation (SSB), and quadrate amplitude modulation (QAM) might be required. The summary of the operational characteristics of linear amplifiers and Class C nonlinear amplifiers is given in Tables 1.2 and 1.3.

1.4 SWITCH-MODE AMPLIFIERS

When the transistor is used as switch, then the amplifier operates in nonlinear mode of operation, and it can be illustrated with the equivalent circuit in Figure 1.6. Classes C, D, E, and F are usually implemented for narrowband-tuned amplifiers when high efficiency is desired with high power. Classes A, B, AB, and C are operated as transconductance amplifiers, and the mode of operation depends on the conduction angle.

TABLE 1.2

Summary of Linear Amplifier Operational Characteristics

Amplifier Class	Mode	Transistor (Q) Operation	Pros	Cons
A	Linear	Always conducting	Most linear, lowest distortion	Poor efficiency
B	Linear	Each device is on half cycle	$\eta_B > \eta_A$	Worse linearity than Class A
AB	Linear	Mid conduction	Improved linearity with respect to Class B	Power dissipation for low signal levels higher than Class B
C	Nonlinear	Each device is on half cycle	High P_o	Inherent harmonics

TABLE 1.3

Summary of Linear Amplifier Performance Parameters

Amplifier Class	Max Efficiency $\eta_{max}\,[\%]$	Normalized RF Output Power $\dfrac{P_{o,max}}{V_{dc}^2/2R_L}$	Normalized $V_{ds,max}\dfrac{V_m}{V_{dc}}$	Normalized $I_{d,max}\dfrac{I_m}{I_{dc}}$	Power Capability $\dfrac{P_{o,max}}{V_m I_m}$
A	50	1	2	2	0.125
B	78.5	1	2	$\pi = 3.14$	0.125
C	86 ($\theta = 71°$)	1	2	3.9	0.11

RF, radio frequency.

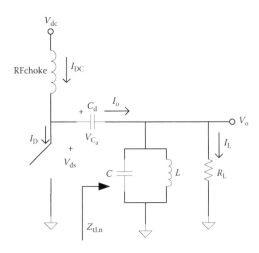

FIGURE 1.6   Equivalent circuit representation of nonlinear amplifier mode of operation.

TABLE 1.4

Summary of Switch-Mode Amplifier Operational Characteristics

Amplifier Class	Max Efficiency η_{max} [%]	Normalized RF Output Power $\dfrac{P_{o,max}}{V_{dc}^2/2R_L}$	Normalized $V_{ds,max}\dfrac{V_m}{V_{dc}}$	Normalized $I_{d,max}\dfrac{I_m}{I_{dc}}$	Power Capability $\dfrac{P_{o,max}}{V_m I_m}$
D	100	$16/\pi^2 = 1.624$	2	$\pi/2 = 1.57$	$1/\pi = 0.318$
E	100	$4/(1 + \pi^2/4) = 1.154$	3.6	2.86	0.098
F	100	$16/\pi^2 = 1.624$	2	$\pi = 3.14$	$1/2\pi = 0.318$

TABLE 1.5

Summary of Switch-Mode Amplifier Performance Parameters

Amplifier Class	Mode	Transistor (Q) Operation	Pros	Cons
D	Switch mode	Q_1 and Q_2 switched on/off alternately	Max efficiency and best power	Device parasitics are issue at high frequencies
E	Switch-mode	Transistor is switched on/off	Max efficiency, no loss due to parasitics	High-voltage stress on transistor
F	Switch mode	Transistor is switched on/off	Max efficiency and no harmonic power delivered	Power loss due to discharge of output capacitance
S	Switch mode	Q_1 and Q_2 are switched on/off with modulated signal	Wider dynamic range and high efficiency	Upper frequency range is limited

In switch-mode amplifiers such as Classes D, E, and F, the active device is intentionally driven into saturation region, and it is operated as a switch rather than a current source unlike Class A, AB, B, or C amplifiers as shown in Figure 1.4. In theory, power dissipation in the transistor can be totally eliminated, and hence, 100% efficiency can be achieved for switch-mode amplifiers.

The summary of some of the switch-mode amplifier performance parameters including efficiency, normalized RF power, normalized maximum drain voltage swing, and power capability are given in Table 1.4. Table 1.5 compares each amplifier class based on transistor operation and application and gives the advantages and disadvantages of each class.

1.5 POWER TRANSISTORS

The selection of a transistor that will be used for amplification in RF amplifier is very critical, because it can affect the performance of the amplifier parameters including efficiency, dissipation, power delivery, stability, linearity, etc. Once the transistor is

TABLE 1.6
Typical Parameter Values for Semiconductor Materials

RF High-Power Material	μ [cm²/Vs]	ε_r	$E_{g\ [eV]}$	Thermal Conductivity [W/cmK]	E_{br} [MV/cm]	JM = $E_{br}v_{sat}/2\pi$	T_{max} [°C]
Si	1350	11.8	1.1	1.3	0.3	1.0	300
GaAs	8500	13.1	1.42	0.46	0.4	2.7	300
SiC	700	9.7	3.26	4.9	3.0	20	600
GaN	1200 (Bulk) 2000 (2DEG)	9.0	3.39	1.7	3.3	27.5	700

RF, radio frequency.

selected for the corresponding amplifier topology, size of the transistor, die placement, bond pads, bonding of the wires, and lead connections will determine the layout of the amplifier and the thermal management of the system. The most commonly used RF and microwave power devices for commercial purposes are based on silicon (Si), gallium arsenide (GaAs), and its compounds. There is an intense research in the development of high-power density devices using materials that have a wide bandgap such as silicon carbide (SiC) and gallium nitride (GaN). Fundamentally, the device performance is determined by several parameters including material energy bandgap, breakdown field, electrons and holes transport properties, thermal conductivity, saturated electron velocity, and conductivity. The typical values of these parameters for various types of semiconductor materials are given in Table 1.6.

The power device family tree can be simplified and is shown in Figure 1.7.

1.5.1 BJT AND HBTs

A typical npn bipolar junction transistor (BJT) is formed out of silicon and the emitter and collector regions that are implanted with donors. There are differences in the characteristics of field effect transistor (FET) and heterojunction bipolar transistor (HBT) devices. The main difference between BJT and HBT is the introduction of heterojunction at the emitter–base interface in the HBT device [3]. This can be illustrated in Figure 1.8.

FET is a planar device, whereas HBT is a vertical device. The HBT device is an enhanced version of conventional BJTs, as a result of the exploitation of heterostructure junctions [4]. Unlike conventional BJTs, in HBTs the bandgap difference between the emitter and the base materials results in higher common emitter gain. Base sheet resistance is lower than that in ordinary BJTs, and the resulting operating frequency is accordingly higher [5].

1.5.2 FETs

The FET family includes a variety of structures, among which are metal-semiconductor field-effect transistors (MESFETs), MOSFETs, HEMTS, etc. They typically consist

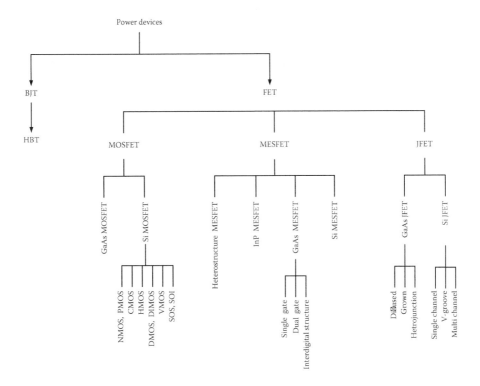

FIGURE 1.7 Family tree for power devices.

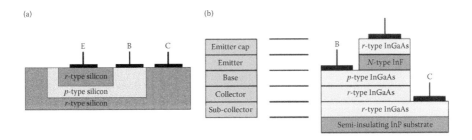

FIGURE 1.8 Typical layer structure of (a) silicon BJT and (b) InP/InGaAs HBT transistor.

of a conductive channel accessed by two ohmic contacts, acting as source (S) and drain (D) terminals, respectively. The third terminal, the gate (G), forms a rectifying junction with the channel or a metal oxide semiconductor (MOS) structure. A simplified structure of a metal–semiconductor n-type FET is depicted in Figure 1.9 [5]. FET devices ideally do not draw current through the gate terminal, unlike the BJTs, which conversely require a significant base current, thus simplifying the biasing arrangement. FET devices exhibit a negative temperature coefficient, resulting

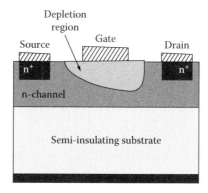

FIGURE 1.9   Simplified structure of FET.

TABLE 1.7
RF Power Transistors and Their Applications and Frequency of Operations

RF Transistor	Drain BV [V]	Frequency [GHz]	Major Applications
RF power FET	65	0.001–0.4	VHF power amplifier
GaAs MesFET	16–22, 60	1–30	Radar, satellite, defense
SiC MesFET	100	0.5–2.3	Base station
GaN MesFET	160	1–30	Replacement for GaAs
Si LDMOS (FET)	65	0.5–2	Base station
Si VDMOS (FET)	65–1200	0.001–0.5	High power amplifiers, FM broadcasting, and magnetic resonance imaging

BV, breakdown voltage; RF, radio frequency; HF, high frequency; LDMOS, laterally diffused MOSFET; MESFET, metal-semiconductor field-effect transistor; VHF, very high frequency.

in a decreasing drain current as the temperature increases. This prevents thermal runaway and allows multiple FETs to be connected in parallel without ballasting, a useful property if a corporate or combined device concept has to be adopted for high-power amplifier design.

RF MOSFET power transistors and their major applications and frequency of operation are given in Table 1.7.

1.5.2.1   MOSFETs

MOSFETs are widely used in RF power amplifier applications, and their parameters are identified by manufacturers at different static and dynamic conditions. Therefore, each MOSFET device has been manufactured with different characteristics. The designer selects the appropriate device for the specific circuit under consideration. One of the standard ways commonly used by designers for selection of right MOSFET device is called figure of merit (FOM) [1]. There are different types of FOMs that are used. FOM in its simplest form compares the gate charge, Q_g, against R_{dsON}. The multiplication of gate charge and drain to source on resistance relates to a certain device technology as it can be related to the required Q_g and R_{dsON} to achieve the right scale for MOSFET. The challenge is the relation between Q_g and R_{dsON} because MOSFET has inherent trade-offs between ON resistance and gate charge i.e., the lower the *12* dsON, the higher the gate charge will be. In device design, this is trade-off between conduction loss versus switching loss. The new generation MOSFETs are manufactured to have an improved FOM [2,6–8]. The comparison of FOM on MOSFETs manufactured with different processes can be illustrated on planar MOSFET structure and trench MOSFET structure. MOSFET with trench structure has seven times better FOM versus planar structure as shown in Figure 1.10.

Two variations of the trench power MOSFET are shown Figure 1.11. The trench technology has the advantage of higher cell density but is more difficult to manufacture than the planar device.

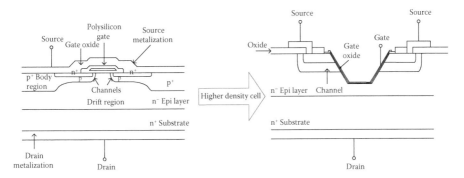

FIGURE 1.10   FOM comparison of planar and trench MOSFET structures.

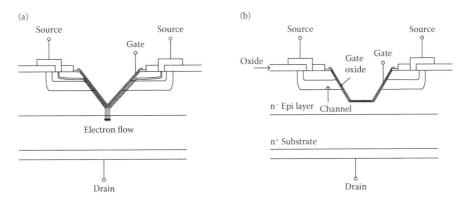

FIGURE 1.11   Trench MOSFET. (a) Current crowding in V-Groove trench MOSFET and (b) truncated V-Groove.

1.6 PASSIVE DEVICES

Passive components used in the design of amplifiers are inductors, capacitors, and resistors. Their size, ratings, and packages vary depending on the operational parameters of the amplifier. Most of the time, capacitors and resistors need to be purchased as off-the-shelf part, whereas inductors can be designed by the amplifier designer.

1.6.1 INDUCTOR

Inductors can be implemented as discrete component or distributed element depending on the frequency of application. If a current flows through a wire wound, a flux is produced through each turn as a result of magnetic flux density as shown in Figure 1.12. The relation between the flux density and flux through each turn can be represented as

$$\Psi = \int \bar{B} \cdot d\bar{s}. \tag{1.4}$$

If there are N turns, then we define the flux linkage as

$$\lambda = N\Psi = N \int \bar{B} \cdot d\bar{s}. \tag{1.5}$$

Inductance is defined as the ratio of flux linkage to current flowing through the windings as defined by

$$L = \frac{\lambda}{I} = \frac{N\Psi}{I}. \tag{1.6}$$

The inductance defined by Equation 1.6 is also known as self-inductance of the core that is formed by the windings. The core can be an air core or a magnetic core.

Inductors can be formed as air-core inductors or magnetic-core inductors depending on the application. When air-core inductors are formed through windings and operated at high frequency (HF), the inductor presents high-frequency

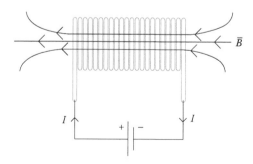

FIGURE 1.12   Flux through each turn.

characteristics. This includes winding resistance and distributed capacitance effects between each turn as shown in Figure 1.13.

The high-frequency model of the inductor can be illustrated with its equivalent circuit, as shown in Figure 1.14.

As a result, the inductor will act as an inductor up to a certain frequency and then gets into resonance and exhibits capacitive effects after resonance frequency. It can be shown that the series equivalent circuit for the high-frequency circuit can be obtained, such as the one shown in Figure 1.15 [9].

where

$$Z = R_s + jX_s, \tag{1.7}$$

where

$$R_s = \frac{R}{\left(1 - \omega^2 L C_s\right)^2 + \left(\omega R C_s\right)^2} \tag{1.8}$$

and

$$X_s = \frac{\omega\left(L - R^2 C_s\right) - \omega^3 L^2 C_s}{\left(1 - \omega^2 L C_s\right)^2 + \left(\omega R C_s\right)^2}. \tag{1.9}$$

The resonance frequency is found when $X_s = 0$ as

$$f_r = \frac{1}{2\pi}\sqrt{\frac{L - R^2 C_s}{L^2 C_s}}. \tag{1.10}$$

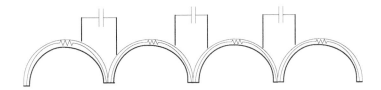

FIGURE 1.13 High-frequency effects of RF inductor.

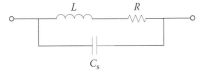

FIGURE 1.14 High-frequency model of RF air-core inductor.

FIGURE 1.15 Equivalent series circuit.

The quality factor is obtained from

$$Q = \frac{|X_s|}{R_s}.$$ (1.11)

C_s in Figure 1.14 is the capacitance including the effects of distributed capacitance of the inductor and is given as

$$C_s = \frac{2\pi\varepsilon_0 da N^2}{l_W}.$$ (1.12)

1.6.1.1   Air-Core Inductor Design

In practice, inductors can be implemented as air-core or toroidal inductors using magnetic cores as illustrated in Figure 1.16.

For an air-core solenoidal inductor given in Figure 1.16, the inductance can be calculated using the relation:

$$L = \frac{d^2 N^2}{18d + 40l} \quad [\mu H].$$ (1.13)

In this equation, L is given as inductance in [μH], d is the coil inner diameter in inches [in.], l is the coil length in inches [in.], and N is the number of turns of the coil. The formula given in Equation 1.13 can be extended to include the spacing between each turn of the air coil inductor. Then, Equation 1.13 can be modified as

$$L = \frac{d^2 N^2}{18d + 40\left(Na + (N-1)s\right)} \quad [\mu H].$$ (1.14)

In Equation 1.14, a represents the wire diameter in inches, and s represents the spacing in inches between each turn.

1.6.1.2   Magnetic-Core Inductor Design

In several RF applications, it may be required to have larger inductance values in space-restricted areas. One solution to increase the inductance value for an air-core

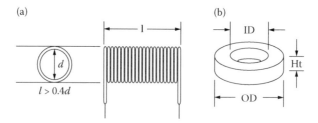

FIGURE 1.16   (a) Air-core inductor and (b) toroidal inductor.

inductor is to increase the number of turns. However, this increases the size of the air-core inductor. This challenge can be overcome by using magnetic cores. Another advantage of using toroidal cores is also keeping the flux within the core as shown in Figure 1.17. This provides self-shielding. In air-core inductor design, air is used as a nonmagnetic material to wind the wire around it. When air is replaced with a magnetic material such as toroidal core, the inductance of the formed inductor can be calculated using

$$L = \frac{4\pi N^2 \mu_i A_{Tc}}{l_e} \quad [\text{nH}]. \tag{1.15}$$

In Equation 1.15, L is the inductance in nanohenries [nH], N is the number of turns, μ_i is the initial permeability, A_{Tc} is the total cross-sectional area of the core in cm^2, and l_e is the effective length of the core in cm. The effective length of the core, l_e, is defined as

$$l_e = \frac{\pi(\text{od} - \text{id})}{\ln(\text{od}/\text{id})} \quad [\text{cm}], \tag{1.16}$$

where od is the outside and id is the inside diameters of the core in cm. The total cross-sectional area of the core, A_{Tc}, is defined as

$$A_{Tc} = \frac{1}{2}(\text{od} - \text{id}) \times h \times n \quad [\text{cm}^2], \tag{1.17}$$

where h refers to the thickness of the core in cm, and n is used to define the number of stacked cores.

It is not uncommon to have the information about the inductance index of the core on its data sheet. If the inductance index is given, then Equation 1.15 can be modified as

$$L = N^2 A_L \quad [\text{nH}], \tag{1.18}$$

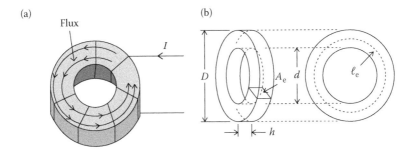

FIGURE 1.17   (a) Magnetic flux in toroidal magnetic core and (b) geometry of toroid.

where A_L is the inductance index in nanohenries/turn². The maximum operational flux density for toroidal core is calculated from

$$B_{op} = \frac{V_{rms} \times 10^8}{4.44 \, fNA_{Tc}} \quad \text{[Gauss]}. \tag{1.19}$$

In Equation 1.19, B_{op} is the magnetic-flux density in Gauss, V_{rms} is the maximum rms voltage across the inductor in volts, f is the frequency in Hertz, N is the number of turns, and A_{Tc} is the total cross-sectional area of the core in cm². The proper design of toroidal core inductor requires operational voltage of the inductor and the required inductance value. This helps to identify the right material for the inductor design to prevent saturation.

1.6.1.3 Planar Inductor Design

In high-frequency applications, planar-type inductors such as spiral inductors can be a good choice to reduce the impact of parasitic effect. Spiral-type planar inductors are widely used in the design of power amplifiers, oscillators, microwave switches, combiners, splitters, etc.

The inductance value of the spiral inductors at the HF range can be determined using the quasi-static method proposed by Greenhouse with a good level of accuracy.

The method proposed by Greenhouse takes into account self-coupling and mutual coupling between each trace. The layout of the two conductors that is used in the inductance calculation is illustrated in Figure 1.18a. GMD is the geometric mean distance between two conductors, and AMD represents the arithmetic mean distance between two conductors. The total inductance of the configuration of the spiral inductor is

$$L_T = L_0 + \Sigma M, \tag{1.20}$$

where L_T is the total inductance, L_0 is the sum of the self-inductances, and ΣM is the sum of the total mutual inductances. The application of the formulation given by Equation 1.20 can be demonstrated for the spiral inductor illustrated in Figure 1.18b and c as

$$L_T = L_1 + L_2 + L_3 + L_4 + L_5 - 2\left(M_{13} + M_{24} + M_{35}\right) + 2M_{15}. \tag{1.21}$$

The general relations that can be used in the algorithm for the spiral inductance calculation then become

$$L_i = 0.0002 l_i \left[\ln\left(2\frac{l_i}{\text{GMD}} \right) - 1.25 + \frac{\text{AMD}}{l_i} + \frac{\mu}{4}T \right] \tag{1.22}$$

$$M_{ij} = 0.0002 l_i Q_i. \tag{1.23}$$

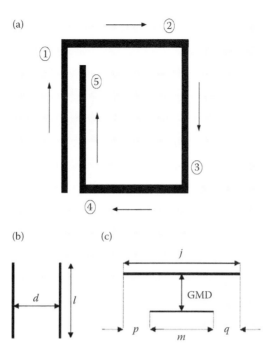

FIGURE 1.18 (a) Rectangular spiral inductor, (b) layout of current filaments, and (c) two-parallel filament geometry.

$$\ln\left(GMD_i\right) = \ln(d) - \frac{1}{12\left(\dfrac{d}{w}\right)^2} - \frac{1}{60\left(\dfrac{d}{w}\right)^4} - \frac{1}{168\left(\dfrac{d}{w}\right)^6} - \frac{1}{360\left(\dfrac{d}{w}\right)^8} - \cdots \qquad (1.24)$$

$$Q_i = \ln\left[\frac{l_i}{GMD_i} + \left(1 + \left(\frac{l_i}{GMD_i}\right)^2\right)^{0.5}\right] - \left(1 + \left(\frac{GMD_i}{l_i}\right)^2\right)^{0.5} + \frac{GMD_i}{l_i} \qquad (1.25)$$

$$AMD = w + t, \qquad (1.26)$$

where L_i is the self-inductance of the segment i, M_{ij} is the mutual inductance between segments i and j, l_i is the length of segment l_i, μ is the permeability of the conductor, T is the frequency correction factor, d is the distance between conductor filaments, w is the width of the conductor, t is the thickness of the conductor, Q_i is the mutual inductance parameter of segment i, GMD_i is the geometric distance of segment i, and AMD is the arithmetic mean distance.

1.6.2 Capacitor

The high-frequency mode of the capacitor is given in Figure 1.19.

From Figure 1.19, it is also clear that nonideal capacitor has resonances due to its high-frequency characteristics. The high-frequency model of the capacitor has parasitic components such as lead inductance, L; conductor loss, R_s; and dielectric loss, R_d, which only become relevant at high frequencies. The characteristics of the capacitor can be obtained by finding the equivalent impedance as

$$Z = (j\omega L_s + R_s) + \left(\frac{1}{G_d + j\omega C}\right) = \frac{R_s G_d^2 + (\omega C)^2 + G_d}{G_d^2 + (\omega C)^2} + j\frac{\omega L G_d^2 + L\omega(\omega C)^2 - \omega C}{G_d^2 + (\omega C)^2}.$$

(1.27)

The Equation 1.27 can be expressed as

$$Z = R_s + jX_s,$$

(1.28)

where

$$R_s = \frac{R_s G_d^2 + (\omega C)^2 + G_d}{G_d^2 + (\omega C)^2} \quad \text{and} \quad X_s = \frac{\omega L G_d^2 + L\omega(\omega C)^2 - \omega C}{G_d^2 + (\omega C)^2}.$$

(1.29)

Impedance given in Equation 1.28 can be converted to admittance as

$$Y = Z^{-1} = \frac{R_s}{R_s^2 + X_s^2} + j\frac{-X_s}{R_s^2 + X_s^2} = G + jB$$

(1.30)

or

$$Y = \frac{\left(R_s G_d^2 + (\omega C)^2 + G_d\right)\left(G_d^2 + (\omega C)^2\right)}{\left(R_s G_d^2 + (\omega C)^2 + G_d\right)^2 + \left(\omega L G_d^2 + L\omega(\omega C)^2 - \omega C\right)^2}$$

$$+ j\frac{\left(\omega C - \omega L G_d^2 - L\omega(\omega C)^2\right)\left(G_d^2 + (\omega C)^2\right)}{\left(R_s G_d^2 + (\omega C)^2 + G_d\right)^2 + \left(\omega L G_d^2 + L\omega(\omega C)^2 - \omega C\right)^2}.$$

(1.31)

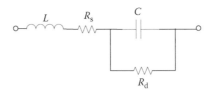

FIGURE 1.19 High-frequency model of capacitor.

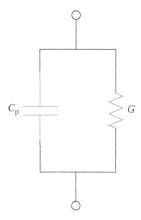

FIGURE 1.20 Equivalent parallel circuit.

Then, the capacitor can be represented by a parallel equivalent circuit as shown in Figure 1.20 where

$$G = \frac{\left(R_s G_d^2 + (\omega C)^2 + G_d\right)\left(G_d^2 + (\omega C)^2\right)}{\left(R_s G_d^2 + (\omega C)^2 + G_d\right)^2 + \left(\omega L G_d^2 + L\omega(\omega C)^2 - \omega C\right)^2}$$ (1.32)

$$B = \frac{\left(\omega C - \omega L G_d^2 - L\omega(\omega C)^2\right)\left(G_d^2 + (\omega C)^2\right)}{\left(R_s G_d^2 + (\omega C)^2 + G_d\right)^2 + \left(\omega L G_d^2 + L\omega(\omega C)^2 - \omega C\right)^2}.$$ (1.33)

The resonance frequency for the circuit shown in Figure 1.14 is found when $B = 0$ as

$$f_r = \frac{1}{2\pi}\sqrt{\frac{R^2 C - L}{R^2 C^2 L}}.$$ (1.34)

The quality factor for the parallel network is then obtained from

$$Q = \frac{|B|}{G} = \frac{R_p}{|X_p|}.$$ (1.35)

1.6.3 RESISTOR

Resistor parasitic inductance is significant for the high-frequency applications. With the increase in the system frequency, the voltage drop across the resistor increases due to the increase in impedance. The type of the resistor whose behavior is frequency dependent is called nonideal resistor. Figure 1.21 is the representation of the impedance model of a resistor. The equivalent circuit of the resistor includes parasitic

(a)   (b)

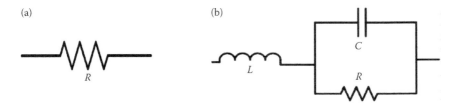

FIGURE 1.21 Representation of the resistor equivalent model. (a) Ideal resistor and (b) equivalent nonideal resistor.

capacitance, lead inductance, and the resistor value. There are several ways of resistor construction. The most common types of resistors are composed of carbon and thin film. Figure 1.22 [10] represents the impedance behavior of the nonideal resistor versus frequency. The breakpoint of the frequency where the resistor acts as a parasitic capacitance and lead inductance can be expressed by Equations 1.36 and 1.37.

When a nonideal resistor acts as a capacitor,

$$f = \frac{1}{2\pi R C_{\text{parasitic}}}. \tag{1.36}$$

When a nonideal resistor acts as an inductor,

$$f = \frac{1}{2\pi \sqrt{L_{\text{lead}} C_{\text{parasitic}}}}. \tag{1.37}$$

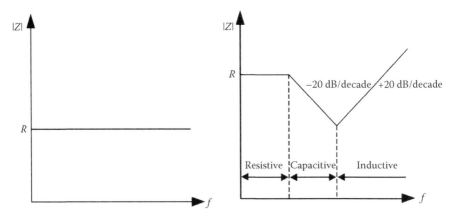

FIGURE 1.22 Characteristics of the resistor with frequency. (a) Impedance of the ideal resistor and (b) nonideal behavior of resistor.

REFERENCES

1. S. Cripps, *RF Power Amplifiers for Wireless Communication*, 2nd ed., Artech House, Norwood, MA, 2006.
2. A. Eroglu, *Introduction to RF Power Amplifier Design and Simulation*, 1st ed., CRC Press, Boca Raton, FL, 2015.
3. U. K. Mishra, Fellow IEEE, L. Shen, T. E. Kazior, and Y.-F. Wu, GaN-based RF power devices and amplifiers, *Proceedings of the IEEE*, Vol. 96, No. 2, pp. 287–305, 2008.
4. M. Golio, *RF and Microwave Semiconductor Device Handbook*, CRC Press, Boca Raton, FL, 2003.
5. S. M. Sze, *Semiconductor Devices Physics and Technology*, John Wiley & Sons, Ltd, New York, 2001.
6. G. Deboy, M. Marz, J.-P. Stengl, H. Strack, J. Tihanyi, and H. Weber, A new generation of high voltage MOSFETs breaks the limit line of silicon, IEDM'98 Electron Devices Meeting, IEDM'98, Technical Digest, pp. 683–685, 1998.
7. G. Sabui and Z. J. Shen, On the feasibility of further improving Figure of Merits (FOM) of low voltage power MOSFETs, in *Proceedings of the 26th International Symposium on Power Semiconductor Devices and IC's*, Waikoloa, HI, June 15–19, 2014.
8. S. Xu et al., NexFET: A new power device, in *Proceedings of International Electron Devices Meeting*, pp. 1–4, 2009.
9. A. Eroglu, *RF Circuit Design Techniques for MF-UHF Applications*, 1st ed., CRC Press, Boca Raton, FL, 2013.
10. C. R. Paul, *Introduction to Electromagnetic Compatibility*, 2nd ed., Wiley-Interscience, Hoboken, NJ, 2006.

2 Two-Port Parameters

2.1 INTRODUCTION

Network parameters allow engineers to determine overall circuit performance without knowing the internal structure. They carry great importance in analysis and design of devices and components. Network parameters provide mathematical tools for designers to model and characterize devices by establishing relationships between voltages and currents. It is possible to theoretically calculate loss, power delivered, reflection coefficient, voltage and current gains, and several other critical parameters with the use of network analysis. Hence, it is necessary to understand and utilize network parameters in radio frequency (RF)/microwave device and component design to have a better performance.

2.2 NETWORK PARAMETERS

The analysis of network parameters can be explained using two-port networks. The two-port network shown in Figure 2.1 is described by a set of four independent parameters, which can be related to voltage and current at any ports of the network. As a result, two-port networks can be treated as a black box modeled by the relationships between the four variables. There exist six different ways to describe the relationships between these variables depending on which two of the four variables are given, whereas the other two can always be derived. All voltages and currents are complex variables and represented by phasors containing both magnitude and phase. Two-port networks are characterized using two-port network parameters such as Z-impedance, Y-admittance, h-hybrid, and $ABCD$. They are usually expressed in matrix notation, and they establish relationships between the following parameters: input voltage V_1, output voltage V_2, input current I_1, and output current I_2. High-frequency networks are characterized by S-parameters.

2.2.1 Z-IMPEDANCE PARAMETERS

The voltages are represented in terms of currents through Z-parameters as follows:

$$V_1 = Z_{11}I_1 + Z_{12}I_2 \tag{2.1}$$

$$V_2 = Z_{21}I_1 + Z_{22}I_2 \tag{2.2}$$

In matrix form, Equations 2.1 and 2.2 can be combined and written as

$$\begin{bmatrix} V_1 \\ V_2 \end{bmatrix} = \begin{bmatrix} Z_{11} & Z_{12} \\ Z_{21} & Z_{22} \end{bmatrix} \begin{bmatrix} I_1 \\ I_2 \end{bmatrix} \tag{2.3}$$

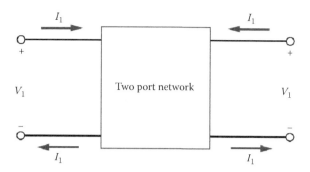

FIGURE 2.1 Two-port network representation.

The Z-parameters for a two-port network are defined as

$$Z_{11} = \frac{V_1}{I_1}\bigg|_{I_2=0} \qquad Z_{12} = \frac{V_1}{I_2}\bigg|_{I_1=0}$$

$$Z_{21} = \frac{V_2}{I_1}\bigg|_{I_2=0} \qquad Z_{22} = \frac{V_2}{I_2}\bigg|_{I_1=0} \qquad (2.4)$$

The formulation in Equation 2.4 can be generalized for N-port network as

$$Z_{nm} = \frac{V_n}{I_m}\bigg|_{I_k=0(k\neq m)} \qquad (2.5)$$

Z_{nm} is the input impedance seen looking into port n, when all other ports are open circuited. In other words, Z_{nm} is the transfer impedance between ports n and m when all other ports are open. It can be shown that for reciprocal networks

$$Z_{nm} = Z_{mn} \qquad (2.6)$$

2.2.2 Y-ADMITTANCE PARAMETERS

The currents are related to voltages through Y-parameters as follows:

$$I_1 = Y_{11}V_1 + Y_{12}V_2 \qquad (2.7)$$

$$I_2 = Y_{21}V_1 + Y_{22}V_2 \qquad (2.8)$$

In matrix form, Equations 2.7 and 2.8 can be written as

$$\begin{bmatrix} I_1 \\ I_2 \end{bmatrix} = \begin{bmatrix} Y_{11} & Y_{12} \\ Y_{21} & Y_{22} \end{bmatrix} \begin{bmatrix} V_1 \\ V_2 \end{bmatrix} \qquad (2.9)$$

The *Y*-parameters in Equation 2.9 can be defined as

$$Y_{11} = \frac{I_1}{V_1}\bigg|_{V_2=0} \qquad Y_{12} = \frac{I_1}{V_2}\bigg|_{V_1=0}$$

$$Y_{21} = \frac{I_2}{V_1}\bigg|_{V_2=0} \qquad Y_{22} = \frac{I_2}{V_2}\bigg|_{V_1=0} \qquad (2.10)$$

The formulation in Equation 2.10 can be generalized for *N*-port network as

$$Y_{nm} = \frac{I_n}{V_m}\bigg|_{V_k=0\,(k\neq m)} \qquad (2.11)$$

Y_{nm} is the input admittance seen looking into port *n* when all other ports are short-circuited. In other words, Y_{nm} is the transfer admittance between ports *n* and *m* when all other ports are short. It can be shown that

$$Y_{nm} = Y_{mn} \qquad (2.12)$$

In addition, it can be further proven that the impedance and admittance matrices are related through

$$[Z] = [Y]^{-1} \qquad (2.13)$$

or

$$[Y] = [Z]^{-1} \qquad (2.14)$$

2.2.3 ABCD PARAMETERS

ABCD parameters relate voltages to currents in the following form for two-port networks.

$$V_1 = AV_1 - BI_2 \qquad (2.15)$$

$$I_1 = CV_1 - DI_2 \qquad (2.16)$$

which can be put in matrix form as

$$\begin{bmatrix} V_1 \\ I_1 \end{bmatrix} = \begin{bmatrix} A & B \\ C & D \end{bmatrix} \begin{bmatrix} V_1 \\ -I_2 \end{bmatrix} \qquad (2.17)$$

ABCD parameters in Equation 2.17 are defined as

$$A = \frac{V_1}{V_2}\bigg|_{I_2=0} \qquad B = \frac{V_1}{-I_2}\bigg|_{V_2=0}$$

$$C = \frac{I_1}{V_2}\bigg|_{I_2=0} \qquad D = \frac{I_1}{-I_2}\bigg|_{V_2=0} \qquad (2.18)$$

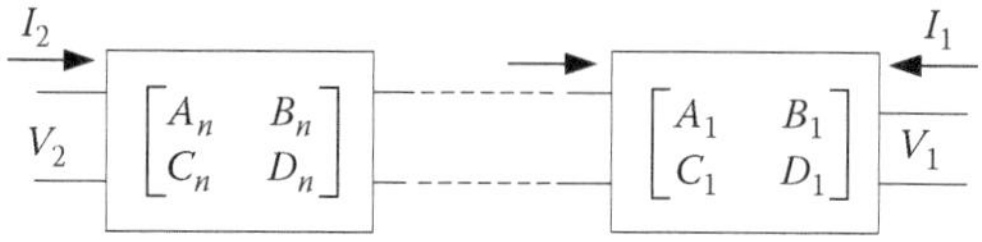

FIGURE 2.2 *ABCD* parameter of cascaded networks.

When network is reciprocal, it can be shown that

$$AD - BC = 1 \tag{2.19}$$

$A = D$ for symmetrical network. *ABCD* parameters are useful in finding voltage or current gain of component or overall gain of a network. One of the great advantages of *ABCD* parameters is their use when networks or components are cascaded. When this condition exists, overall *ABCD* parameter of the network simply becomes the matrix product of individual network or components as given by Equation 2.20. This can be generalized for *N*-port network shown in Figure 2.2 as

$$\left\{ \begin{matrix} v_1 \\ i_1 \end{matrix} \right\} = \left(\begin{bmatrix} A_1 & B_1 \\ C_1 & D_1 \end{bmatrix} \cdots \begin{bmatrix} A_n & B_n \\ C_n & D_n \end{bmatrix} \right) \left\{ \begin{matrix} v_2 \\ -i_2 \end{matrix} \right\} \tag{2.20}$$

2.2.4 *h*-HYBRID PARAMETERS

Hybrid parameters relate voltages and currents in a two-port network as

$$V_1 = h_{11}I_1 + h_{12}V_2 \tag{2.21}$$

$$I_2 = h_{21}I_1 + h_{22}V_2 \tag{2.22}$$

Equations 2.21 and 2.22 can be put in a matrix form as

$$\begin{bmatrix} V_1 \\ I_2 \end{bmatrix} = \begin{bmatrix} h_{11} & h_{12} \\ h_{21} & h_{22} \end{bmatrix} \begin{bmatrix} I_1 \\ V_2 \end{bmatrix} \tag{2.23}$$

The hybrid parameters in Equation 2.23 can be found from

$$h_{11} = \left. \frac{V_1}{I_1} \right|_{V_2=0} \qquad h_{12} = \left. \frac{V_1}{V_2} \right|_{I_1=0}$$
$$\qquad\qquad\qquad\qquad\qquad , \tag{2.24}$$
$$h_{21} = \left. \frac{I_2}{I_1} \right|_{V_2=0} \qquad h_{22} = \left. \frac{I_2}{V_2} \right|_{I_1=0}$$

Hybrid parameters are preferred for components such as transistors and transformers since they can be measured with ease in practice.

Example 2.1

Find (a) impedance, (b) admittance, (c) *ABCD*, and (d) hybrid parameters of the T-network given in Figure 2.3.

Solution:

a. *Z*-parameters are found with application of Equation 2.4 by opening all the other ports except the measurement port. This leads to

$$Z_{11} = \frac{V_1}{I_1}\bigg|_{I_2=0} = Z_A + Z_C \qquad Z_{21} = \frac{V_2}{I_1}\bigg|_{I_2=0} = Z_C$$

$$Z_{12} = \frac{V_1}{I_2}\bigg|_{I_1=0} = \frac{V_2}{I_2}\frac{Z_C}{Z_B+Z_C} = (Z_B+Z_C)\frac{Z_C}{Z_B+Z_C} = Z_C \qquad Z_{22} = \frac{V_2}{I_2}\bigg|_{I_1=0} = Z_B + Z_C{}'$$

The *Z*-matrix is then constructed as

$$Z = \begin{bmatrix} Z_A + Z_C & (Z_B+Z_C)\dfrac{Z_C}{Z_B+Z_C} = Z_C \\[2mm] (Z_B+Z_C)\dfrac{Z_C}{Z_B+Z_C} = Z_C & Z_B + Z_C \end{bmatrix}$$

b. *Y*-parameters are found from Equation 2.10 by shorting all the other ports except the measurement port. Y_{11} and Y_{21} are found when port 2 is shorted as

$$Y_{11} = \frac{I_1}{V_1}\bigg|_{V_2=0} \rightarrow I_1 = \frac{V_1}{Z_A + (Z_B//Z_C)} = V_1\left(\frac{Z_B+Z_C}{Z_A Z_B + Z_A Z_C + Z_B Z_C}\right)$$

$$\rightarrow Y_{11} = \left(\frac{Z_B+Z_C}{Z_A Z_B + Z_A Z_C + Z_B Z_C}\right)$$

$$Y_{21} = \frac{I_2}{V_1}\bigg|_{V_2=0} \rightarrow I_2 = \frac{-V_1}{\left(Z_A + (Z_B//Z_C)\right)}\frac{Z_C}{(Z_C+Z_B)} \rightarrow Y_{21} = \left(\frac{-Z_C}{Z_A Z_B + Z_A Z_C + Z_B Z_C}\right)$$

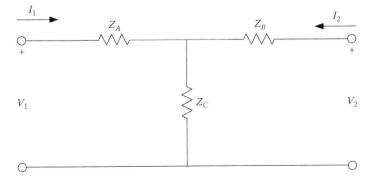

FIGURE 2.3   T-network configuration.

Similarly, Y_{12} and Y_{22} are found when port 1 is shorted as

$$Y_{12} = \frac{I_1}{V_2}\bigg|_{V_1=0} \rightarrow I_1 = \frac{-V_2}{\left(Z_B + (Z_A // Z_C)\right)} \frac{Z_C}{(Z_A + Z_C)} \rightarrow Y_{12} = \left(\frac{-Z_C}{Z_A Z_B + Z_A Z_C + Z_B Z_C}\right)$$

$$Y_{22} = \frac{I_2}{V_2}\bigg|_{V_1=0} \rightarrow I_2 = \frac{V_2}{Z_B + (Z_A // Z_C)} = V_1\left(\frac{Z_A + Z_C}{Z_A Z_B + Z_A Z_C + Z_B Z_C}\right)$$

$$\rightarrow Y_{22} = \left(\frac{Z_A + Z_C}{Z_A Z_B + Z_A Z_C + Z_B Z_C}\right)$$

Y-parameters can also be found by just inverting the Z-matrix given by Equation 2.14 as

$$[Y] = [Z]^{-1} = \frac{1}{(Z_A Z_B + Z_A Z_C + Z_B Z_C)}\begin{bmatrix} Z_B + Z_C & -Z_C \\ -Z_C & Z_A + Z_C \end{bmatrix}$$

So, the Y-matrix for T-network is then

$$Y = \begin{bmatrix} \left(\dfrac{Z_B + Z_C}{Z_A Z_B + Z_A Z_C + Z_B Z_C}\right) & \left(\dfrac{-Z_C}{Z_A Z_B + Z_A Z_C + Z_B Z_C}\right) \\ \left(\dfrac{-Z_C}{Z_A Z_B + Z_A Z_C + Z_B Z_C}\right) & \left(\dfrac{Z_A + Z_C}{Z_A Z_B + Z_A Z_C + Z_B Z_C}\right) \end{bmatrix}$$

As seen from the results of part (a) and (b), the network is reciprocal because

$$Z_{12} = Z_{21} \text{ and } Y_{12} = Y_{21}$$

c. Hybrid parameters are found using Equation 2.24. Parameters h_{11} and h_{21} are obtained when port 2 is shorted as

$$h_{11} = \frac{V_1}{I_1}\bigg|_{V_2=0} \rightarrow V_1 = I_1\left(Z_A + (Z_B // Z_C)\right) = I_1\left(\frac{Z_A Z_B + Z_A Z_C + Z_B Z_C}{Z_B + Z_C}\right)$$

$$\rightarrow h_{11} = \left(\frac{Z_A Z_B + Z_A Z_C + Z_B Z_C}{Z_B + Z_C}\right)$$

and

$$h_{21} = \frac{I_2}{I_1}\bigg|_{V_2=0} \rightarrow I_2 = -I_1\left(\frac{Z_C}{Z_B + Z_C}\right) \rightarrow h_{21} = -\left(\frac{Z_C}{Z_B + Z_C}\right)$$

Parameters h_{12} and h_{22} are obtained when port 1 is open-circuited as

$$h_{12} = \frac{V_1}{V_1}\bigg|_{I_1=0} \rightarrow V_1 = V_2\left(\frac{Z_C}{Z_B + Z_C}\right) \rightarrow h_{12} = \left(\frac{Z_C}{Z_B + Z_C}\right)$$

and

$$h_{22} = \frac{I_2}{V_2}\bigg|_{I_1=0} \rightarrow I_2 = V_2\left(\frac{1}{Z_B + Z_C}\right) \rightarrow h_{22} = \left(\frac{1}{Z_B + Z_C}\right)$$

The hybrid matrix for T-network now can be constructed as

$$h = \begin{bmatrix} \left(\dfrac{Z_A Z_B + Z_A Z_C + Z_B Z_C}{Z_B + Z_C}\right) & \left(\dfrac{Z_C}{Z_B + Z_C}\right) \\ -\left(\dfrac{Z_C}{Z_B + Z_C}\right) & \left(\dfrac{1}{Z_B + Z_C}\right) \end{bmatrix}$$

d. *ABCD* parameters are found using Equations 2.1 through 2.18. Parameters A and C are determined when port 2 is open-circuited as

$$A = \frac{V_1}{V_2}\bigg|_{I_2=0} \rightarrow V_2 = \frac{Z_C}{Z_C + Z_A} V_1 \rightarrow A = \left(\frac{Z_C + Z_A}{Z_C}\right)$$

and

$$C = \frac{I_1}{V_2}\bigg|_{I_2=0} \rightarrow I_1 = V_2\left(\frac{1}{Z_C}\right) \rightarrow C = \left(\frac{1}{Z_C}\right)$$

Parameters B and D are determined when port 2 is short-circuited as

$$B = \frac{V_1}{-I_2}\bigg|_{V_2=0} \rightarrow I_2 = \frac{-V_1}{Z_A + (Z_B//Z_C)} \frac{Z_C}{(Z_B + Z_C)} \rightarrow B = \left(\frac{Z_A Z_B + Z_A Z_C + Z_B Z_C}{Z_C}\right)$$

and

$$D = \frac{-I_1}{I_2}\bigg|_{V_2=0} \rightarrow I_2 = -I_1\left(\frac{Z_C}{Z_B + Z_C}\right) \rightarrow D = \left(\frac{Z_B + Z_C}{Z_C}\right)$$

So, the *ABCD* matrix is found as

$$ABCD = \begin{bmatrix} \left(\dfrac{Z_C + Z_A}{Z_C}\right) & \left(\dfrac{Z_A Z_B + Z_A Z_C + Z_B Z_C}{Z_C}\right) \\ \left(\dfrac{1}{Z_C}\right) & \left(\dfrac{Z_B + Z_C}{Z_C}\right) \end{bmatrix}$$

It can proved from the results obtained that Z, Y, h, and *ABCD* parameters are related using the relations given in Table 2.1.

TABLE 2.1

Network Parameter Conversion Table

	[Z]	[Y]	[ABCD]	[h]
[Z]	$\begin{bmatrix} z_{11} & z_{12} \\ z_{21} & z_{22} \end{bmatrix}$	$\begin{bmatrix} \dfrac{y_{22}}{\Delta_Y} & \dfrac{-y_{12}}{\Delta_Y} \\ \dfrac{-y_{21}}{\Delta_Y} & \dfrac{y_{11}}{\Delta_Y} \end{bmatrix}$	$\begin{bmatrix} \dfrac{A}{C} & \dfrac{\Delta_T}{C} \\ \dfrac{1}{C} & \dfrac{D}{C} \end{bmatrix}$	$\begin{bmatrix} \dfrac{\Delta_H}{h_{22}} & \dfrac{h_{12}}{h_{22}} \\ \dfrac{-h_{21}}{h_{22}} & \dfrac{1}{h_{22}} \end{bmatrix}$
[Y]	$\begin{bmatrix} \dfrac{z_{22}}{\Delta_Z} & \dfrac{-z_{12}}{\Delta_Z} \\ \dfrac{-z_{21}}{\Delta_Z} & \dfrac{z_{11}}{\Delta_Z} \end{bmatrix}$	$\begin{bmatrix} y_{11} & y_{12} \\ y_{21} & y_{22} \end{bmatrix}$	$\begin{bmatrix} \dfrac{D}{B} & \dfrac{-\Delta_T}{B} \\ \dfrac{1}{B} & \dfrac{A}{B} \end{bmatrix}$	$\begin{bmatrix} \dfrac{1}{h_{11}} & \dfrac{-h_{12}}{h_{11}} \\ \dfrac{h_{21}}{h_{11}} & \dfrac{\Delta_H}{h_{11}} \end{bmatrix}$
[ABCD]	$\begin{bmatrix} \dfrac{z_{11}}{z_{21}} & \dfrac{\Delta_Z}{z_{21}} \\ \dfrac{1}{z_{21}} & \dfrac{z_{22}}{z_{21}} \end{bmatrix}$	$\begin{bmatrix} \dfrac{-y_{22}}{y_{21}} & \dfrac{-1}{y_{21}} \\ \dfrac{-\Delta_y}{y_{21}} & \dfrac{-y_{11}}{y_{21}} \end{bmatrix}$	$\begin{bmatrix} A & B \\ C & D \end{bmatrix}$	$\begin{bmatrix} \dfrac{-\Delta_H}{h_{21}} & \dfrac{-h_{11}}{h_{21}} \\ \dfrac{-h_{22}}{h_{21}} & \dfrac{-1}{h_{21}} \end{bmatrix}$
[h]	$\begin{bmatrix} \dfrac{\Delta_Z}{z_{22}} & \dfrac{z_{12}}{z_{22}} \\ \dfrac{-z_{21}}{z_{22}} & \dfrac{1}{z_{22}} \end{bmatrix}$	$\begin{bmatrix} \dfrac{1}{y_{11}} & \dfrac{-y_{12}}{y_{11}} \\ \dfrac{y_{21}}{y_{11}} & \dfrac{\Delta_Y}{y_{11}} \end{bmatrix}$	$\begin{bmatrix} \dfrac{B}{D} & \dfrac{\Delta_T}{D} \\ -\dfrac{1}{D} & -\dfrac{C}{D} \end{bmatrix}$	$\begin{bmatrix} h_{11} & h_{12} \\ h_{21} & h_{22} \end{bmatrix}$

2.3 NETWORK CONNECTIONS

Networks and components in engineering applications can be connected in different ways to perform certain tasks. Commonly used network connection methods are series, parallel, and cascade connections. Series connection of two networks is shown in Figure 2.4a. Because the networks are connected in series, currents are same and voltages are added across ports of the network to find the overall voltage at the ports of the combined network. This can be represented by impedance matrices as

$$[Z] = [Z^x] + [Z^y] = \begin{bmatrix} Z_{11}^x & Z_{12}^x \\ Z_{21}^x & Z_{22}^x \end{bmatrix} + \begin{bmatrix} Z_{11}^y & Z_{12}^y \\ Z_{21}^y & Z_{22}^y \end{bmatrix} = \begin{bmatrix} Z_{11}^x + Z_{11}^y & Z_{12}^x + Z_{12}^y \\ Z_{21}^x + Z_{21}^y & Z_{22}^x + Z_{22}^y \end{bmatrix}$$

(2.25)

So,

$$\begin{bmatrix} V_1 \\ V_2 \end{bmatrix} = \begin{bmatrix} Z_{11}^x + Z_{11}^y & Z_{12}^x + Z_{12}^y \\ Z_{21}^x + Z_{21}^y & Z_{22}^x + Z_{22}^y \end{bmatrix} \begin{bmatrix} I_1 \\ I_2 \end{bmatrix}$$

(2.26)

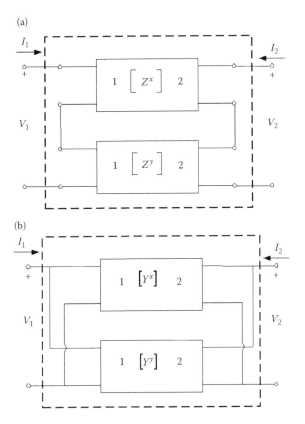

FIGURE 2.4   (a) Series connection of two-port networks and (b) parallel connection of two-port networks.

Parallel connection of two-port networks is illustrated in Figure 2.4b. In parallel-connected networks, voltages are same across ports, and currents are added to find the overall current flowing at the ports of the combined network. This can be represented by Y matrices as

$$[Y]=[Y^x]+[Y^y]=\begin{bmatrix} Y_{11}^x & Y_{12}^x \\ Y_{21}^x & Y_{22}^x \end{bmatrix}+\begin{bmatrix} Y_{11}^y & Y_{12}^y \\ Z_{21}^y & Y_{22}^y \end{bmatrix}=\begin{bmatrix} Y_{11}^x+Y_{11}^y & Y_{12}^x+Y_{12}^y \\ Y_{21}^x+Y_{21}^y & Y_{22}^x+Y_{22}^y \end{bmatrix}$$

(2.27)

As a result,

$$\begin{bmatrix} I_1 \\ I_2 \end{bmatrix}=\begin{bmatrix} Y_{11}^x+Y_{11}^y & Y_{12}^x+Y_{12}^y \\ Y_{21}^x+Y_{21}^y & Y_{22}^x+Y_{22}^y \end{bmatrix}\begin{bmatrix} V_1 \\ V_2 \end{bmatrix}$$

(2.28)

The cascade connection of two-port networks is shown in Figure 2.5. In cascade connection, the magnitude of the current flowing at the output of the first network is

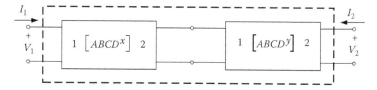

FIGURE 2.5 Cascade connection of two-port networks.

equal to current at the input port of the second network. The voltages at the output of the first network is also equal to the voltage across the input of the second network. This can be represented by using $ABCD$ matrices as

$$[ABCD] = [ABCD^x][ABCD^y] = \begin{bmatrix} A^x & B^x \\ C^x & D^x \end{bmatrix} \begin{bmatrix} A^y & B^y \\ C^y & D^y \end{bmatrix}$$

$$= \begin{bmatrix} A^x A^y + B^x C^y & A^x B^y + B^x D^y \\ C^x A^y + D^x C^y & C^x B^y + D^x D^y \end{bmatrix} \tag{2.29}$$

Example 2.2

Consider RF amplifier given in Figure 2.6. It has feedback network for stability, input, and output matching networks. The transistor used is NPN BJT and its characteristic parameters are given by $r_{BE} = 400\,\Omega$, $r_{CE} = 70\,\text{k}\Omega$, $C_{BE} = 15\,\text{pF}$, and $C_{BC} = 2\,\text{pF}$, and $g_m = 0.2\,\text{S}$. Find the voltage and current gain of this amplifier when $L = 2\,\text{nH}$, $C = 12\,\text{pF}$, $l = 5\,\text{cm}$, and $v_p = 0.65\,c$.

Solution: The high-frequency characteristics of the transistor is modeled using the hybrid parameters given by

$$h_{11} = h_{ie} = \frac{r_{BE}}{1 + j\omega(C_{BE} + C_{BC})r_{BE}} \tag{2.30}$$

$$h_{12} = h_{re} = \frac{j\omega C_{BC} r_{BE}}{1 + j\omega(C_{BE} + C_{BC})r_{BE}} \tag{2.31}$$

$$h_{21} = h_{fe} = \frac{r_{BE}(g_m - j\omega C_{BC})}{1 + j\omega(C_{BE} + C_{BC})r_{BE}} \tag{2.32}$$

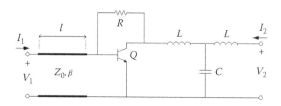

FIGURE 2.6 RF amplifier analysis by network parameters.

$$h_{22} = h_{oe} = \frac{\left[1 + j\omega\left(C_{BE} + C_{BC}\right)r_{BE}\right] + \left[\left(1 + r_{BE}g_m + j\omega C_{BE}r_{BE}\right)\right]r_{CE}}{1 + j\omega\left(C_{BE} + C_{BC}\right)r_{BE}}$$ (2.33)

The amplifier network shown in Figure 2.6 is a combination of four networks that are connected in parallel and cascade. Overall network has to be first partitioned for analysis. This can be demonstrated as shown in Figure 2.7.

In the partitioned amplifier circuit, networks N_2 and N_3 are connected in parallel as shown in Figure 2.8. Then, the parallel-connected network, Y, can be represented by admittance matrix. The admittance matrix of network 3 is

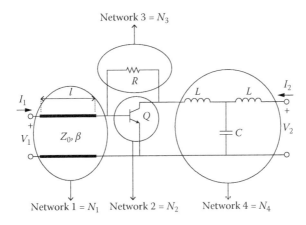

FIGURE 2.7 Partition of amplifier circuit for network analysis.

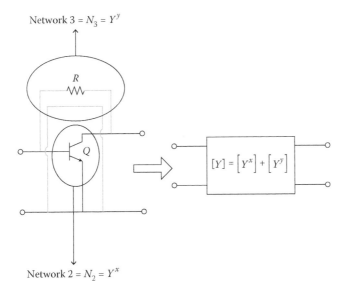

FIGURE 2.8 Illustration of parallel connection between networks 2 and 3.

$$Y^y = \begin{bmatrix} \dfrac{1}{R} & -\dfrac{1}{R} \\[2mm] -\dfrac{1}{R} & \dfrac{1}{R} \end{bmatrix} \tag{2.34}$$

The admittance matrix for the transistor can be now obtained by using the network conversion Table 2.1 since hybrid parameters for it are available. From Table 2.1,

$$Y^x = \begin{bmatrix} \dfrac{1}{h_{11}} & -\dfrac{h_{12}}{h_{11}} \\[2mm] \dfrac{h_{21}}{h_{11}} & \dfrac{\Delta h}{h_{11}} \end{bmatrix} \tag{2.35}$$

Then, overall admittance matrix is found as

$$[Y] = \begin{bmatrix} Y^x \end{bmatrix} + \begin{bmatrix} Y^y \end{bmatrix} = \begin{bmatrix} \dfrac{1}{R} + \dfrac{1}{h_{11}} & -\dfrac{1}{R} - \dfrac{h_{12}}{h_{11}} \\[2mm] -\dfrac{1}{R} + \dfrac{h_{21}}{h_{11}} & \dfrac{1}{R} + \dfrac{\Delta h}{h_{11}} \end{bmatrix} \tag{2.36}$$

where Δ is used for determinant of the corresponding matrix. At this point, it is now clearer that networks 1, Y and 4 are cascaded. We need to determine that $ABCD$ matrix of each network in this connection as shown in Figure 2.9. The first step is then to convert the admittance matrix in Equation 1.36 to $ABCD$ parameter using the conversion table. The conversion table gives the relation as

$$ABCD^Y = \begin{bmatrix} \dfrac{Y_{22}}{Y_{21}} & -\dfrac{1}{Y_{21}} \\[2mm] \dfrac{\Delta Y}{Y_{21}} & \dfrac{Y_{11}}{Y_{21}} \end{bmatrix} \tag{2.37}$$

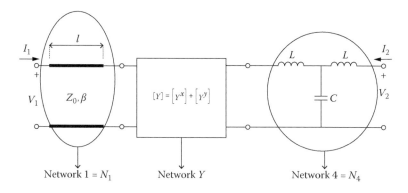

FIGURE 2.9   Cascade connection of final circuit.

ABCD matrices for transmission line network 1 and matching network 4 are obtained as

$$ABCD^{N_1} = \begin{bmatrix} \cos(\beta l) & jZ_0 \sin(\beta l) \\ \dfrac{j \sin(\beta l)}{Z_0} & \cos(\beta l) \end{bmatrix} \tag{2.38}$$

$$ABCD^{N_4} = \begin{bmatrix} 1 - \omega^2 LC & j\omega L\left(2 - \omega^2 LC\right) \\ j\omega C & 1 - \omega^2 LC \end{bmatrix}_a \tag{2.39}$$

The overall *ABCD* parameter of the combined network shown in Figure 2.9 is

$$ABCD = ABCD^{N_1}\left(ABCD^Y\right)ABCD^{N_4} \tag{2.40}$$

or

MATLAB® Script for network analysis of RF amplifier

```
Zo=50;
l=0.05;
L=2e-9;
C=12e-12;
rbe=400;
rce=70e3;
Cbe=15e-12;
Cbc=2e-12;
gm=0.2;
VGain=zeros(5,150);
IGain=zeros(5,150);
freq=zeros(1,150);
R=[200 300 500 1000 10000];

for i=1:5
for t=1:150;

f=10^((t+20)/20);
freq(t)=f;
lambda=0.65*3e8/(f);
bet=(2*pi)/lambda;
w=2*pi*f;
N1=[cos(bet*l) 1j*Zo*sin(bet*l);1j*(1/Zo)*sin(bet*l)
cos(bet*l)];
Y1=[1/R(i) -1/R(i);-1/R(i) 1/R(i)];
k=(1+1j*w*rbe*(Cbc+Cbe));
h=[(rbe/k) (1j*w*rbe*Cbc)/k;(rbe.*(gm-1j*w*Cbc))/k
((1/rce)+(1j*w*Cbc*(1+gm*rbe+1j*w*Cbe*rbe)/k))];
Y2=[1/h(1,1) -h(1,2)/h(1,1);h(2,1)/h(1,1) det(h)/h(1,1)];
Y=Y1+Y2;
N23=[-Y(2,2)/Y(2,1) -1/Y(2,1);(det(Y)/Y(2,1)) -Y(1,1)/Y(2,1)];
N4=[(1-(w^2)*L*C) (2j*w*L-1j*(w^3)*L^2*C);1j*(w*C)
(1-(w^2)*L*C)];
```

```
NT=N1*N23*N4;
VGain(i,t)=20*log10(abs(1/NT(1,1)));
IGain(i,t)=20*log10(abs(-1/NT(2,2)));

end
end

figure
semilogx(freq,(IGain))
axis([10^4 10^9 20 50]);
ylabel('I_{Gain} (I_2/I_1) (dB)');
xlabel('Freq (Hz)');
legend('R=20Ohm','R=30Ohm','R=50Ohm','R=100Ohm','R=10000
Ohm')
figure
semilogx(freq,(VGain))
axis([10^4 10^9 20 80]);
ylabel('V_{Gain} (V_2/V_1) (dB)');
xlabel('Freq (Hz)');
legend('R=20Ohm','R=30Ohm','R=50Ohm','R=100Ohm',
'R=10000Ohm')
```

$$ABCD = \begin{bmatrix} \cos(\beta l) & jZ_0\sin(\beta l) \\ \dfrac{j\sin(\beta l)}{Z_0} & \cos(\beta l) \end{bmatrix} \begin{bmatrix} \dfrac{Y_{22}}{Y_{21}} & -\dfrac{1}{Y_{21}} \\ \dfrac{\Delta Y}{Y_{21}} & \dfrac{Y_{11}}{Y_{21}} \end{bmatrix} \begin{bmatrix} 1-\omega^2 LC & j\omega L(2-\omega^2 LC) \\ j\omega C & 1-\omega^2 LC \end{bmatrix}$$

(2.41)

Voltage and current gains from $ABCD$ parameters are found using

$$V_{\text{Gain}} = 20\log\left(\left|\frac{1}{A}\right|\right)[\text{dB}]$$

(2.42)

$$I_{\text{Gain}} = 20\log\left(\left|\frac{1}{D}\right|\right)[\text{dB}]$$

(2.43)

MATLAB script has been written to obtain the voltage and current gains. The script that can be used for analysis of any other similarly constructed amplifier network is given for reference. Voltage and current gains are obtained by MATLAB versus various feedback resistor values, and the frequencies are shown in Figures 2.10 and 2.11. This type of analysis gives the designer the ability to study the effect of several parameters on output response in an amplifier circuit including feedback, matching networks, and parameters of the transistor.

2.4   S-SCATTERING PARAMETERS

Scattering parameters are used to characterize RF/microwave devices and components at high frequencies [1–3]. Specifically, they are used to define the return loss and insertion loss of a component or device.

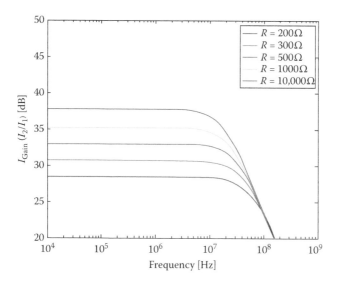

FIGURE 2.10 Current gain of RF amplifier versus feedback resistor values and frequency.

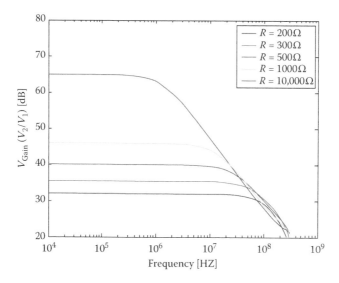

FIGURE 2.11 Voltage gain of RF amplifier versus feedback resistor values and frequency.

2.4.1 ONE-PORT NETWORK

Consider the circuit given in Figure 2.12. The relationship between current and voltage can be written as

$$I = \frac{V_g}{Z_g + Z_L} \tag{2.44}$$

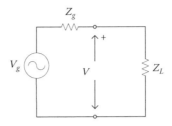

FIGURE 2.12   One-port network for scattering parameter analysis.

and

$$V = \frac{V_g Z_L}{Z_g + Z_L} \tag{2.45}$$

where Z_g is the generator impedance. The incident waves for voltage and current can be obtained when the generator is matched as

$$I_i = \frac{V_g}{Z_g + Z_g^*} = \frac{V_g}{2 \operatorname{Re}\{Z_g\}} \tag{2.46}$$

and

$$V_i = \frac{V_g Z_g^*}{Z_g + Z_g^*} = \frac{V_g Z_g^*}{2 \operatorname{Re}\{Z_g\}} \tag{2.47}$$

Then, the reflected waves are found from

$$I = I_i - I_r \tag{2.48}$$

and

$$V = V_i - V_r \tag{2.49}$$

Substituting Equations 2.44 and 2.46 into Equation 2.48 gives the reflected wave as

$$I_r = I_i - I = \left(\frac{Z_L - Z_g^*}{Z_L + Z_g^*} \right) I_i \tag{2.50}$$

or

$$I_r = S^I I_i \tag{2.51}$$

where

$$S^I = \left(\frac{Z_L - Z_g^*}{Z_L + Z_g^*} \right) \tag{2.52}$$

is the scattering matrix for current. Similar analysis can be done to find reflected voltage wave by substituting Equations 2.45 and 2.47 into Equation 2.49 as

$$V_r = V_i - V = \frac{Z_g}{Z_g^*}\left(\frac{Z_L - Z_g^*}{Z_L + Z_g^*}\right)V_i \qquad (2.53)$$

or

$$V_r = \frac{Z_g}{Z_g^*}S^I V_i = S^V V_i \qquad (2.54)$$

where

$$S^V = \frac{Z_g}{Z_g^*}S^I \qquad (2.55)$$

is the scattering matrix for voltage. It can also be shown that

$$V_i = Z_g^* I_i \qquad (2.56)$$

$$V_r = Z_g I_r \qquad (2.57)$$

When generator impedance is pure real, $Z_g = R_g$, then

$$S^I = S^V = \left(\frac{Z_L - R_g}{Z_L + R_g}\right)_i \qquad (2.58)$$

2.4.2 N-PORT NETWORK

The analysis described in the previous section can be extended to *N*-port network shown in Figure 2.13. The analysis is based on the assumption that generators are independent of each other. Hence, *Z* generator matrix has no cross coupling terms, and it can be expressed as a diagonal matrix.

$$[Z_g] = \begin{bmatrix} Z_{g1} & 0 & \cdots & 0 \\ 0 & Z_{g2} & \cdots & 0 \\ \vdots & \vdots & & \vdots \\ 0 & 0 & \cdots & Z_{gn} \end{bmatrix} \qquad (2.59)$$

From Equations 2.48 and 2.49, the incident and reflected waves are related to the actual voltage and current values as

$$[I] = [I_i] - [I_r] \qquad (2.60)$$

$$[V] = [V_i] + [V_r] \qquad (2.61)$$

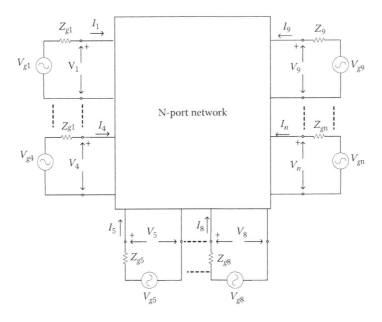

FIGURE 2.13   N-port network for scattering analysis.

From Equations 2.56 and 2.57, the incident and reflected components can be related through

$$[V_i] = \left[Z_g^*\right][I_i] \tag{2.62}$$

$$[V_r] = \left[Z_g\right][I_r] \tag{2.63}$$

similar to one-port case as derived before. For N-port network, Z-parameters can be obtained as

$$[V] = [Z][I] \tag{2.64}$$

Using Equations 2.59 through 2.64, we can obtain

$$[V_r] = [V] - [V_i] = [Z][I] - \left[Z_g^*\right][I_i] \tag{2.65}$$

Equation 2.65 can be also expressed as

$$\left[Z_g\right][I_r] = [Z][I] - \left[Z_g^*\right][I_i] = [Z]\left(\left[I_i\right] - \left[I_r\right]\right) - \left[Z_g^*\right][I_i] \tag{2.66}$$

and simplified to

$$\left([Z] + \left[Z_g\right]\right)[I_r] = \left([Z] - \left[Z_g^*\right]\right)[I_i] \tag{2.67}$$

Equation 2.67 can be put in the following form

$$[I_r] = ([Z]+[Z_g])^{-1}([Z]-[Z_g^*])[I_i]$$ (2.68)

From Equation 2.52, the scattering matrix for current for N-port network is equal to

$$S' = ([Z]+[Z_g])^{-1}([Z]-[Z_g^*])$$ (2.69)

Then, Equation 2.68 can be expressed as

$$[I_r] = [S'][I_i]$$ (2.70)

For N-port network, Y-parameters for short circuit case can be obtained similarly as

$$[I] = [Y][V]$$ (2.71)

It can also be shown that

$$[V_r] = -([Y]+[Y_g])^{-1}([Y]-[Y_g^*])[V_i]$$ (2.72)

or

$$[V_r] = [S^V][V_i]$$ (2.73)

where

$$S^V = -([Y]+[Y_g])^{-1}([Y]-[Y_g^*])$$ (2.74)

Example 2.3

Consider a transistor network that is represented as a two-port network and connected between source and load. It is assumed that the generator or source and load impedances are equal and given to be R_g. Transistor is represented by the following Z-parameters as shown in Figure 2.14. Find current scattering matrix, S'.

$$[Z] = \begin{bmatrix} Z_i & Z_r \\ Z_f & Z_o \end{bmatrix}$$

Solution: From Equation 2.69, the scattering matrix for current is

$$S' = ([Z]+[Z_g])^{-1}([Z]-[Z_g^*])$$ (2.75)

The generator Z_g-matrix is

$$[Z_g] = [Z_g^*] = \begin{bmatrix} R_g & 0 \\ 0 & R_g \end{bmatrix}$$ (2.76)

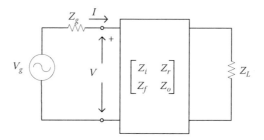

FIGURE 2.14   Two-port transistor network.

Then,

$$[Z]+[Z_g^*] = \begin{bmatrix} Z_i & Z_r \\ Z_f & Z_o \end{bmatrix} + \begin{bmatrix} R_g & 0 \\ 0 & R_g \end{bmatrix} = \begin{bmatrix} Z_i+R_g & 0 \\ 0 & Z_o+R_g \end{bmatrix} \quad (2.77)$$

The inverse of the matrix in Equation 2.77 is

$$\left[\left([Z]+[Z_g^*]\right) \right]^{-1} = \frac{1}{\left| [Z]+[Z_g^*] \right|} \left(\left[[Z]+[Z_g^*] \right]^C \right)^T \quad (2.78)$$

$\left| [Z]+[Z_g^*] \right|$ is the determinant of $[Z]+[Z_g^*]$ and is calculated as

$$\left| [Z]+[Z_g^*] \right| = \left(Z_i+R_g \right)\left(Z_o+R_g \right) - Z_r Z_f \quad (2.79)$$

$\left[[Z]+[Z_g^*] \right]^C$ is the cofactor matrix for $[Z]+[Z_g^*]$ and calculated as

$$\left([Z]+[Z_g^*] \right)^C = \begin{bmatrix} Z_o+R_g & -Z_f \\ -Z_r & Z_i+R_g \end{bmatrix} \quad (2.80)$$

Then,

$$\left[\left([Z]+[Z_g^*] \right)^C \right]^T = \begin{bmatrix} Z_o+R_g & -Z_r \\ -Z_f & Z_i+R_g \end{bmatrix} \quad (2.81)$$

Hence, the inverse of the matrix from Equations 2.77 to 2.81 is equal to

$$\left[\left([Z]+[Z_g^*]\right) \right]^{-1} = \frac{1}{\left(\left(Z_i+R_g \right)\left(Z_o+R_g \right) - Z_r Z_f \right)} \begin{bmatrix} Z_o+R_g & -Z_r \\ -Z_f & Z_i+R_g \end{bmatrix} \quad (2.82)$$

Then, from Equation 2.75

$$S' = \left(\left[Z\right]+\left[Z_g\right]\right)^{-1}\left(\left[Z\right]-\left[Z_g\right]\right)$$

$$= \left(\left[\frac{1}{\left(\left(Z_i+R_g\right)\left(Z_o+R_g\right)-Z_rZ_f\right)}\begin{bmatrix} Z_o+R_g & -Z_r \\ -Z_f & Z_i+R_g \end{bmatrix}\right]-\begin{bmatrix} Z_i-R_g & Z_r \\ Z_f & Z_o-R_g \end{bmatrix}\right)$$

(2.83)

which can be simplified to

$$S' = \left(\left[Z\right]+\left[Z_g\right]\right)^{-1}\left(\left[Z\right]-\left[Z_g\right]\right)$$

$$= \begin{pmatrix} \left(Z_o+R_g\right)\left(Z_i-R_g\right)-Z_rZ_f & 2Z_rR_g \\ 2Z_fR_g & \left(Z_i+R_g\right)\left(Z_o-R_g\right)-Z_rZ_f \end{pmatrix}$$

(2.84)

2.4.3 NORMALIZED SCATTERING PARAMETERS

Normalized scattering parameters can be introduced by a and b for the incident and reflected waves as follows.

$$[a] = \frac{1}{\sqrt{2}}\sqrt{\left(\left[Z_g\right]+\left[Z_g^*\right]\right)}\left[I_i\right]$$

(2.85)

$$[b] = \frac{1}{\sqrt{2}}\sqrt{\left(\left[Z_g\right]+\left[Z_g^*\right]\right)}\left[I_r\right]$$

(2.86)

where

$$\frac{1}{\sqrt{2}}\sqrt{\left(\left[Z_g\right]+\left[Z_g^*\right]\right)} = \sqrt{\text{Re}\left\{Z_g\right\}} = \begin{bmatrix} \sqrt{\text{Re}\left\{Z_{g1}\right\}} & 0 & K & 0 \\ 0 & \sqrt{\text{Re}\left\{Z_{g2}\right\}} & L & 0 \\ M & M & & M \\ 0 & 0 & L & \sqrt{\text{Re}\left\{Z_{gn}\right\}} \end{bmatrix}$$

(2.87)

Substituting Equation 2.68 into Equations 2.85 and 2.86 gives

$$\frac{[b]}{\sqrt{\text{Re}\left\{Z_g\right\}}} = [I_r] = \left[S'\right]\left[I_i\right]$$

(2.88)

or

$$\frac{[b]}{\sqrt{\mathrm{Re}\{Z_{\mathrm{g}}\}}} = [S'] \frac{[a]}{\sqrt{\mathrm{Re}\{Z_{\mathrm{g}}\}}} \tag{2.89}$$

Then, from Equations 2.88 and 2.89,

$$[b] = \sqrt{\mathrm{Re}\{Z_{\mathrm{g}}\}} [S'] [\mathrm{Re}\{Z_{\mathrm{g}}\}]^{-1/2} [a] \tag{2.90}$$

Equation 2.90 can be simplified to

$$[b] = [S][a] \tag{2.91}$$

where

$$[S] = \sqrt{\mathrm{Re}\{Z_{\mathrm{g}}\}} [S'] [\mathrm{Re}\{Z_{\mathrm{g}}\}]^{-1/2} \tag{2.92}$$

and

$$[\mathrm{Re}\{Z_{\mathrm{g}}\}]^{-1/2} = \begin{bmatrix} \dfrac{1}{\sqrt{\mathrm{Re}\{Z_{\mathrm{g1}}\}}} & 0 & K & 0 \\[2ex] 0 & \dfrac{1}{\sqrt{\mathrm{Re}\{Z_{\mathrm{g2}}\}}} & L & 0 \\[2ex] M & M & M & M \\[2ex] 0 & 0 & L & \dfrac{1}{\sqrt{\mathrm{Re}\{Z_{\mathrm{gn}}\}}} \end{bmatrix} \tag{2.93}$$

S matrix in Equation 2.91 is called a normalized scattering matrix. It can be proven that

$$[S'] = [Z_{\mathrm{g}}]^{-1} [S^V] [Z_{\mathrm{g}}^*] \tag{2.94}$$

When the generator or source impedance is real, $Z_{\mathrm{g}} = R_{\mathrm{g}}$, then from Equation 2.94, we obtain

$$[S'] = [S^V] \tag{2.95}$$

In addition, Equations 2.69 and 2.92 take the following form

$$S' = ([Z] + [R_{\mathrm{g}}])^{-1} ([Z] - [R_{\mathrm{g}}]) \tag{2.96}$$

$$[S] = \sqrt{R_{\mathrm{g}}} [S'] [Z_{\mathrm{g}}]^{-1/2} \tag{2.97}$$

From Equations 2.95 and 2.97, we obtain

$$[S] = [S^I] = [S^V] \tag{2.98}$$

S-parameters can be calculated using the two-port network shown in Figure 2.15. In Figure 2.15, the source or generator impedances are given as R_{g1} and R_{g2}. When Equation 2.91 is expanded,

$$\begin{bmatrix} b_1 \\ b_2 \end{bmatrix} = \begin{bmatrix} S_{11} & S_{12} \\ S_{21} & S_{22} \end{bmatrix} \begin{bmatrix} a_1 \\ a_2 \end{bmatrix} \tag{2.99}$$

From Equation 2.99,

$$b_1 = S_{11}a_1 + S_{12}a_2 \tag{2.100}$$

$$b_2 = S_{21}a_1 + S_{22}a_2 \tag{2.101}$$

Hence, S-parameters can be defined from Equations 2.100 and 2.101 as

$$S_{11} = \frac{b_1}{a_1}\bigg|_{a_2=0} \qquad S_{12} = \frac{b_1}{a_2}\bigg|_{a_1=0} \tag{2.102}$$

$$S_{21} = \frac{b_2}{a_1}\bigg|_{a_2=0} \qquad S_{22} = \frac{b_2}{a_2}\bigg|_{a_1=0}$$

From Equation 2.102, the scattering parameters are calculated when $a_1=0$ or $a_2=0$. a represents the incident waves. If Equation 2.85 is reviewed again,

$$[a] = \frac{1}{\sqrt{2}} \sqrt{\left([Z_g] + [Z_g^*] \right)} [I_i] \tag{2.103}$$

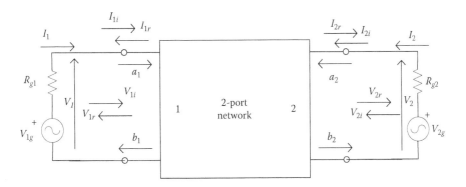

FIGURE 2.15   S-parameters for two-port networks.

a_2 becomes zero when $I_{2i} = 0$. This can be obtained when there is no source connected to port 2, i.e., $V_{2g} = 0$ with existence of source impedance, R_{g2}. From KLV for the second port, we obtain

$$V_2 = -I_2 R_{g2} \text{ or } V_2 + I_2 R_{g2} = 0 \tag{2.104}$$

Substituting Equations 2.60 and 2.61 into Equation 2.104 gives

$$V_2 + I_2 R_{g2} = V_{i2} + V_{2r} + R_{g2}\left(I_{2i} - I_{2r}\right) \tag{2.105}$$

which leads to

$$V_2 + I_2 R_{g2} = I_{2i} R_{g2} + R_{g2} I_{2r} + R_{g2} I_{2i} - I_{2r} R_{g2} \tag{2.106}$$

or

$$V_2 + I_2 R_{g2} = 2 R_{g2} I_{2i} \tag{2.107}$$

From Equation 2.103, when $Z_g = R_g$,

$$[a] = \sqrt{R_g}\left[I_i\right] \tag{2.108}$$

Substituting Equation 2.106 into Equation 2.107 gives

$$V_2 + I_2 R_{g2} = 2\sqrt{R_{g2}}\, a_2 \tag{2.109}$$

Then,

$$a_2 = \frac{V_2 + I_2 R_{g2}}{2\sqrt{R_{g2}}} \tag{2.110}$$

It is then proven that when Equation 2.104 is substituted into Equation 2.110, $a_2 = 0$ as expected. This also requires $I_i = 0$ from Equation 2.109. Then, this shows that there is no reflected current, which is the incident current, I_{2i}, at port 2 owing to the source generator incident wave from port 1.

Similar analysis can be done at port 1 when $a_1 = 0$. Following the same steps, it can be shown that

$$a_1 = \frac{V_1 + I_1 R_{g1}}{2\sqrt{R_{g1}}} \tag{2.111}$$

Reflected waves, b_1 and b_2, can be analyzed the same way using the analysis just presented for the incident waves a_1 and a_2. When no source voltage is connected at port 1, $a_1 = 0$ in the existence of source voltage R_{g1}, we can write

$$V_1 = -I_1 R_{g1} \text{ or } V_1 + I_1 R_{g1} = 0 \tag{2.112}$$

In terms of the reflected and incident voltage and current, we get

$$V_1 - I_1 R_{g1} = V_{1i} + V_{1r} - R_{g1}\left(I_{1i} - I_{1r}\right) \tag{2.113}$$

which leads to

$$V_1 - I_1 R_{g1} = I_{1i} R_{g1} + I_{1r} R_{g1} - I_{1i} R_{g1} + I_{1r} R_{g1} \tag{2.114}$$

or

$$V_1 - I_1 R_{g1} = 2 R_{g1} I_{1r} \tag{2.115}$$

From Equation 2.86, when $Z_g = R_g$,

$$[b] = \sqrt{R_g}\,[I_r] \tag{2.116}$$

Hence, Equation 2.115 can be written as

$$V_1 - I_1 R_{g1} = 2\sqrt{R_{g1}}\, b_1 \tag{2.117}$$

Then,

$$b_1 = \frac{V_1 - I_1 R_{g1}}{2\sqrt{R_{g1}}} \tag{2.118}$$

We can also show that when $a_2 = 0$,

$$b_2 = \frac{V_2 - I_2 R_{g2}}{2\sqrt{R_{g2}}} \tag{2.119}$$

Incident and reflected parameters a and b for N-port network can be written using the results given in Equations 2.110, 2.111, 2.118, and 2.119 for real generator impedance, R_g, as

$$[a] = \frac{1}{2}[R_g]^{-1/2}\left([V] + [R_g][I]\right) \tag{2.120}$$

$$[b] = \frac{1}{2}[R_g]^{-1/2}\left([V] - [R_g][I]\right) \tag{2.121}$$

For an arbitrary impedance, Equations 2.120 and 2.121 can be written as

$$[a] = \frac{1}{2}\left[\mathrm{Re}\{Z_g\}\right]^{-1/2}\left([V] + [Z_g][I]\right) \tag{2.122}$$

$$[b] = \frac{1}{2}\left[\mathrm{Re}\{Z_g\}\right]^{-1/2}\left([V] - [Z_g^*][I]\right) \tag{2.123}$$

Now, since we derived the conditions when a_1 and a_2 are zero, we can expand equations given by Equation 2.102. When $a_2=0$, we can calculate S_{11} and S_{21}. From Equations 2.111, 2.118, 2.108, and 2.106, S_{11} can be expressed as

$$S_{11} = \left.\frac{b_1}{a_1}\right|_{a_2=0} = \left.\frac{\left(\dfrac{V_1 - I_1 R_{g1}}{2\sqrt{R_{g1}}}\right)}{\left(\dfrac{V_1 + I_1 R_{g1}}{2\sqrt{R_{g1}}}\right)}\right|_{I_{2i}=0} = \frac{V_1 - I_1 R_{g1}}{V_1 + I_1 R_{g1}} = \frac{V_{1r}}{V_{1i}} = \frac{\sqrt{R_{1g}} I_{1r}}{\sqrt{R_{1g}} I_{1i}} = \frac{I_{1r}}{I_{1i}} \quad (2.124)$$

or

$$S_{11} = \frac{Z_{11} - R_{g1}}{Z_{11} + R_{g1}} \quad (2.125)$$

In Equation 2.125, S_{11} is the reflection coefficient at port 1 when port 2 is terminated with generator impedance R_{g2}. We express S_{21} using Equations 2.111, 2.119, 2.108, and 2.116 as

$$S_{21} = \left.\frac{b_2}{a_1}\right|_{a_2=0} = \left.\frac{\left(\dfrac{V_2 - I_2 R_{g2}}{2\sqrt{R_{g2}}}\right)}{\left(\dfrac{V_1 + I_1 R_{g1}}{2\sqrt{R_{g1}}}\right)}\right|_{I_{2i}=0} = \frac{\left(V_2 - I_2 R_{g2}\right)\sqrt{R_{g1}}}{\left(V_1 + I_1 R_{g1}\right)\sqrt{R_{g2}}} = \frac{\sqrt{R_{2g}} I_{2r}}{\sqrt{R_{1g}} I_{1i}} \quad (2.126)$$

When $a_2=0$, $V_{2g}=0$, and that results in $V_2=-I_2 Rg_2$ and $V_{1g}=2I_{1i}R_{g1}$; then Equation 2.126 can be written as

$$S_{21} = \left.\frac{b_2}{a_1}\right|_{a_2=0} = -\frac{\sqrt{R_{2g}} I_2}{\sqrt{R_{1g}} \left(V_{1g}/2R_{1g}\right)} = -2\sqrt{R_{1g}}\sqrt{R_{2g}} \frac{I_2}{V_{1g}} = 2\sqrt{\frac{R_{1g}}{R_{2g}}} \frac{V_2}{V_{1g}} = \frac{\left(V_2/\sqrt{R_{2g}}\right)}{\left(\dfrac{1}{2} V_{1g}/\sqrt{R_{1g}}\right)}$$

$$(2.127)$$

As shown from Equation 2.127, S_{21} is the forward transmission gain of the network from port 1 to port 2. Similar procedure can be repeated to derive S_{22} and S_{12} when $a_1=0$. Hence, it can be shown that

$$S_{22} = \left.\frac{b_2}{a_2}\right|_{a_1=0} = \left.\frac{\left(\dfrac{V_2 - I_2 R_{g2}}{2\sqrt{R_{g2}}}\right)}{\left(\dfrac{V_2 + I_2 R_{g2}}{2\sqrt{R_{g2}}}\right)}\right|_{I_{1i}=0} = \frac{V_2 - I_2 R_{g2}}{V_2 + I_2 R_{g2}} = \frac{V_{2r}}{V_{2i}} = \frac{\sqrt{R_{2g}} I_{2r}}{\sqrt{R_{2g}} I_{2i}} = \frac{I_{2r}}{I_{2i}} \quad (2.128)$$

or

$$S_{22} = \frac{Z_{22} - R_{g2}}{Z_{22} + R_{g2}} \qquad (2.129)$$

S_{22} is the reflection coefficient of the output. S_{12} can be obtained as

$$S_{12} = \frac{b_1}{a_2}\bigg|_{a_1=0} = \frac{\left(\dfrac{V_1 - I_1 R_{g1}}{2\sqrt{R_{g1}}}\right)}{\left(\dfrac{V_2 + I_2 R_{g2}}{2\sqrt{R_{g2}}}\right)}\Bigg|_{I_{1i}=0} = \frac{(V_1 - I_1 R_{g1})\sqrt{R_{g2}}}{(V_2 + I_2 R_{g2})\sqrt{R_{g1}}} = \frac{\sqrt{R_{1g}}\,I_{1r}}{\sqrt{R_{2g}}\,I_{2i}} \qquad (2.130)$$

which can be put in the following form

$$S_{12} = \frac{b_1}{a_2}\bigg|_{a_1=0} = -2\sqrt{R_{1g}}\sqrt{R_{2g}}\frac{I_1}{V_{2g}} = 2\sqrt{\frac{R_{2g}}{R_{1g}}}\frac{V_1}{V_{2g}} = \frac{\left(V_1/\sqrt{R_{1g}}\right)}{\left(\dfrac{1}{2}V_{2g}/\sqrt{R_{2g}}\right)} \qquad (2.131)$$

S_{12} is the reverse transmission gain of the network from port 2 to port 1. Overall, S-parameters are found when $a_n = 0$, which means that there is no reflection at that port. This is only possible by matching all the ports except the measurement port. Insertion loss and return loss in terms of S-parameters are defined as

$$\text{Insertion Loss [dB]} = \text{IL [dB]} = 20\log\left(\left|S_{ij}\right|\right),\ i \neq j \qquad (2.132)$$

$$\text{Return Loss [dB]} = \text{RL [dB]} = 20\log\left(\left|S_{ii}\right|\right) \qquad (2.133)$$

Another important parameter that can be defined using S-parameters is the voltage standing wave ratio (VSWR). For instance, VSWR at port 1 is found from

$$\text{VSWR} = \frac{1 - |S_{11}|}{1 + |S_{11}|} \qquad (2.134)$$

The two-port network is reciprocal if

$$S_{21} = S_{12} \qquad (2.135)$$

It can be shown that a network is reciprocal if it is equal to its transpose. This is represented for a two-port network as

$$[S] = [S]^t \qquad (2.136)$$

or

$$\begin{bmatrix} S_{11} & S_{12} \\ S_{21} & S_{22} \end{bmatrix}^t = \begin{bmatrix} S_{11} & S_{21} \\ S_{12} & S_{22} \end{bmatrix} \qquad (2.137)$$

When network is lossless, S-parameters can be used to characterize this feature as

$$[S]^t [S]^* = [U] \qquad (2.138)$$

where $*$ defines complex conjugate of a matrix, and U is the unitary matrix and defined by

$$[U] = \begin{bmatrix} 1 & 0 \\ 0 & 1 \end{bmatrix} \qquad (2.139)$$

Equation 2.138 can be applied for two-port network as

$$[S]^t [S]^* = \begin{bmatrix} \left(|S_{11}|^2 + |S_{21}|^2\right) & \left(S_{11}S_{12}^* + S_{21}S_{22}^*\right) \\ \left(S_{12}S_{11}^* + S_{22}S_{21}^*\right) & \left(|S_{12}|^2 + |S_{22}|^2\right) \end{bmatrix} = \begin{bmatrix} 1 & 0 \\ 0 & 1 \end{bmatrix} \qquad (2.140)$$

We can further show that if a network is lossless and reciprocal, it satisfies

$$|S_{11}|^2 + |S_{21}|^2 = 1 \qquad (2.141)$$

$$S_{11}S_{12}^* + S_{21}S_{22}^* = 0 \qquad (2.142)$$

Example 2.4

Find the characteristic impedance of the T-network given in Figure 2.16 to have no return loss at the input port.

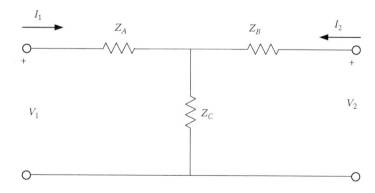

FIGURE 2.16   T-network configuration.

Solution: Scattering parameter for T-network are found from Equation 2.102. From Equation 2.102, S_{11} is equal to

$$S_{11} = \left.\frac{b_1}{a_1}\right|_{a_2=0} = \frac{Z_{in} - Z_o}{Z_{in} + Z_o} \qquad (2.143)$$

where

$$Z_{in} = Z_A + \left[\frac{Z_C(Z_B + Z_o)}{Z_C + (Z_B + Z_o)}\right] \qquad (2.144)$$

No return loss is possible when $S_{11} = 0$. This can be satisfied from Equations 2.143 and 2.144 when

$$Z_o = Z_{in} = Z_A + \left[\frac{Z_C(Z_B + Z_o)}{Z_C + (Z_B + Z_o)}\right] \qquad (2.145)$$

REFERENCES

1. A. Eroglu, *RF Circuit Design Techniques for MF-UHF Applications*, CRC Press, Boca Raton, FL, 2013.
2. G. Matthaei, E. M .T. Jones, and L. Young, *Microwave Filters, Impedance-Matching Networks, and Coupling Structures*, Artech House, Norwood, MA, 1980.
3. A. Eroglu, *Introduction to RF Power Amplifier Design and Simulation*, CRC press, Boca Raton, FL, 2016.

3 Impedance Matching and Resonant Networks

3.1 INTRODUCTION

Radio frequency (RF) power amplifiers consist of several stages as illustrated in Figure 3.1 [1–2]. Impedance matching networks are used to provide the optimum power transfer from one stage to another, so that the energy transfer is maximized. This can be accomplished by having matching networks between the stages. Matching networks can be implemented using distributed or lumped elements based on the frequency of operation and application. Distributed elements are implemented using transmission lines for high-frequency operation where lumped elements are used for lower frequencies. Matching networks, when designed with lumped elements, are implemented using ladder network structure. In the design of matching network, there are several important parameters such as bandwidth and quality factor of the network. These can be investigated with design tools such as Smith chart. Smith chart helps the designer to visualize the performance of the matching network for the operational conditions under consideration. In this chapter, analysis of transmission lines, Smith chart, and design of impedance matching network will be detailed, and several application examples will be given.

3.2 TRANSMISSION LINES

A transmission line is a distributed-parameter network, where voltages and currents can vary in magnitude and phase over the length of the line. Transmission lines usually consist of two parallel conductors that can be represented with a short segment of Δz. This short segment of transmission line can be modeled as a lumped-element circuit as shown in Figure 3.2.

In Figure 3.2, R is the series resistance per unit length for both conductors, $R[\Omega/m]$; L is the series inductance per unit length for both conductors, $L[H/m]$; G is the shunt conductance per unit length, $G[S/m]$; and C represents the shunt capacitance per unit length, $C[F/m]$ in the transmission line. Application of Kirchhoff's Voltage Law (KVL) and Kirchhoff's Current Law (KCL) give

$$v(z,t) - R\Delta z i(z,t) - L\Delta z \frac{\partial i(z,t)}{\partial t} - v(z+\Delta z, t) = 0 \tag{3.1}$$

$$i(z,t) - G\Delta z v(z+\Delta z, t) - C\Delta z \frac{\partial v(z+\Delta z, t)}{\partial t} - i(z+\Delta z, t) = 0. \tag{3.2}$$

Dividing Equations 3.1 and 3.2 by Δz and assuming $\Delta z \rightarrow 0$, we obtain

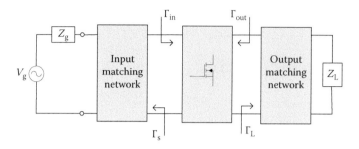

FIGURE 3.1 Matching network implementation for RF power amplifiers.

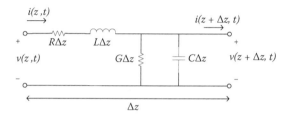

FIGURE 3.2 Short segment of transmission line.

$$\frac{\partial v(z,t)}{\partial z} = -Ri(z,t) - L\frac{\partial i(z,t)}{\partial t}$$ (3.3)

$$\frac{\partial i(z,t)}{\partial z} = -Gv(z,t) - C\frac{\partial v(z,t)}{\partial t}.$$ (3.4)

Equations 3.3 and 3.4 are known as the time-domain form of the transmission line or telegrapher equations. Assuming the sinusoidal steady-state condition with application cosine-based phasors, Equations 3.3 and 3.4 take the following forms:

$$\frac{dV(z)}{dz} = -(R + j\omega L)I(z).$$ (3.5)

$$\frac{dI(z)}{dz} = -(G + j\omega C)V(z).$$ (3.6)

By eliminating either $I(z)$ or $V(z)$ from Equations 3.5 to 3.6, we obtain the wave equations as

$$\frac{d^2V(z)}{dz^2} = -\gamma^2 V(z)$$ (3.7)

$$\frac{d^2I(z)}{dz^2} = -\gamma^2 I(z),$$ (3.8)

where

$$\gamma = \alpha + j\beta = \sqrt{(R + j\omega L)(G + j\omega C)} \tag{3.9}$$

In Equation 3.9, γ is the complex propagation constant, α is attenuation constant, and β is known as the phase constant. In transmission lines, phase velocity is defined as

$$v_p = \frac{\omega}{\beta}. \tag{3.10}$$

Wavelength can be defined using

$$\lambda = \frac{2\pi}{\beta}. \tag{3.11}$$

Traveling wave solutions to the equations obtained in Equations 3.7 and 3.8 are

$$V(z) = V_o^+ e^{-\gamma z} + V_o^- e^{+\gamma z} \tag{3.12}$$

$$I(z) = I_0^+ e^{-\gamma z} + I_0^- e^{+\gamma z}. \tag{3.13}$$

Substitution of Equations 3.12 into 3.5 gives

$$I(z) = \frac{\gamma}{R + j\omega L} \left[V_o^+ e^{-\gamma z} + V_o^- e^{+\gamma z} \right]. \tag{3.14}$$

From Equation 3.14, the characteristic impedance, Z_0, is defined as

$$Z_0 = \frac{R + j\omega L}{\gamma} = \sqrt{\frac{R + j\omega L}{G + j\omega C}}. \tag{3.15}$$

Hence,

$$\frac{V_o^+}{I_0^+} = Z_0 = -\frac{V_o^-}{I_0^-} \tag{3.16}$$

and

$$I(z) = \frac{V_o^+}{Z_0} e^{-\gamma z} - \frac{V_o^-}{Z_0} e^{+\gamma z} \tag{3.17}$$

Using the formulation derived, we can find the voltage and current at any point on the transmission line (Figure 3.3). At the load $z = 0$,

$$V(0) = Z_L I(0) \tag{3.18}$$

$$V_o^+ + V_o^- = \frac{Z_L}{Z_0} \left(V_o^+ - V_o^- \right) \tag{3.19}$$

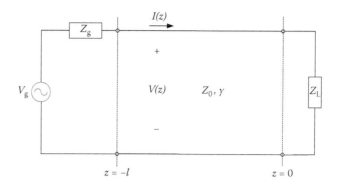

FIGURE 3.3 Finite terminated transmission line.

or

$$V_0^-\left(1+\frac{Z_L}{Z_0}\right)=V_0^+\left(\frac{Z_L}{Z_0}-1\right), \tag{3.20}$$

which leads to

$$\frac{V_0^-}{V_0^+}=\left(\frac{Z_L-Z_0}{Z_L+Z_0}\right). \tag{3.21}$$

We then define Equation 3.21 as the reflection coefficient at the load and express it as

$$\Gamma_L=\left(\frac{Z_L-Z_0}{Z_L+Z_0}\right). \tag{3.22}$$

We can express voltage and current in terms of reflection coefficient as

$$V(z)=V_0^+\left(e^{-\gamma z}+\Gamma_L\,e^{+\gamma z}\right) \tag{3.23}$$

$$I(z)=\frac{1}{Z_0}V_0^+\left(e^{-\gamma z}-\Gamma_L\,e^{+\gamma z}\right). \tag{3.24}$$

We can find the input impedance at any point on the transmission line shown in Figure 3.4 from

$$Z_{in}(z)=\frac{V(z)}{I(z)}. \tag{3.25}$$

We then have

$$Z_{in}(z)=Z_0\frac{\left(e^{-\gamma z}+\Gamma_L\,e^{+\gamma z}\right)}{\left(e^{-\gamma z}-\Gamma_L\,e^{+\gamma z}\right)}, \tag{3.26}$$

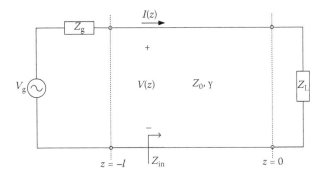

FIGURE 3.4 Input impedance calculation on the transmission line.

which can be expressed as

$$Z_{in}(z) = Z_0 \left(\frac{1 + \left(\dfrac{Z_L - Z_0}{Z_L + Z_0} \right) e^{+2\gamma z}}{1 - \left(\dfrac{Z_L - Z_0}{Z_L + Z_0} \right) e^{+2\gamma z}} \right) = Z_0 \left(\frac{(Z_L + Z_0) + (Z_L - Z_0) e^{+2\gamma z}}{(Z_L + Z_0) - (Z_L - Z_0) e^{+2\gamma z}} \right)$$

$$= Z_0 \left(\frac{(Z_L + Z_0) e^{-\gamma z} + (Z_L - Z_0) e^{+\gamma z}}{(Z_L + Z_0) e^{-\gamma z} - (Z_L - Z_0) e^{+\gamma z}} \right). \tag{3.27}$$

We can rewrite Equation 3.27 as

$$Z_{in}(z) = Z_0 \left(\frac{(Z_L + Z_0) e^{-\gamma z} + (Z_L - Z_0) e^{+\gamma z}}{(Z_L + Z_0) e^{-\gamma z} - (Z_L - Z_0) e^{+\gamma z}} \right) = Z_0 \left(\frac{Z_L \left(e^{+\gamma z} + e^{-\gamma z} \right) - Z_0 \left(e^{+\gamma z} - e^{-\gamma z} \right)}{-Z_L \left(e^{+\gamma z} - e^{-\gamma z} \right) + Z_0 \left(e^{+\gamma z} + e^{-\gamma z} \right)} \right), \tag{3.28}$$

which can also be expressed as

$$Z_{in}(z) = Z_0 \left(\frac{Z_L - Z_0 \tanh(\gamma z)}{Z_0 - Z_L \tanh(\gamma z)} \right) \tag{3.29}$$

At the input when $z = -l$, the impedance can be found from Equation 3.29 as

$$Z_{in}(z) = Z_0 \left(\frac{Z_L + Z_0 \tanh(\gamma l)}{Z_0 + Z_L \tanh(\gamma l)} \right). \tag{3.30}$$

3.2.1 LIMITING CASES FOR TRANSMISSION LINES

There are three cases that can be considered as the limiting case for transmission lines. These are lossless lines, low loss lines, and distortionless lines.

1. Lossless line ($R = G = 0$)

Transmission lines can be considered as lossless when $R = G = 0$. When $R = G = 0$, the defining equations for the transmission lines can be simplified as

$$\gamma = \alpha + j\beta = j\omega\sqrt{LC} \Rightarrow \alpha = 0 \tag{3.31a}$$

$$\beta = \omega\sqrt{LC} \tag{3.31b}$$

$$v_p = \frac{\omega}{\beta} = \frac{1}{\sqrt{LC}} \tag{3.31c}$$

$$Z_0 = \sqrt{\frac{L}{C}} = R_0 + jX_0 \Rightarrow R_0 = \sqrt{\frac{L}{C}}, \ X_0 = 0. \tag{3.31d}$$

2. Low loss line ($R \ll \omega L, G \ll \omega C$)

For low-loss transmission lines, $R \ll \omega L$, $G \ll \omega C$, and defining equations simplify to

$$\gamma = \alpha + j\beta = j\omega\sqrt{LC}\left(1 + \frac{R}{j\omega L}\right)^{1/2}\left(1 + \frac{G}{j\omega C}\right)^{1/2} \tag{3.32a}$$

$$\alpha \cong \frac{1}{2}\left(R\sqrt{\frac{C}{L}} + G\sqrt{\frac{L}{C}}\right) \tag{3.32b}$$

$$\beta \cong \omega\sqrt{LC} \tag{3.32c}$$

$$v_p = \frac{\omega}{\beta} \cong \frac{1}{\sqrt{LC}} \tag{3.32d}$$

$$Z = R_0 + jX_0 = \sqrt{\frac{L}{C}}\left(1 + \frac{R}{j\omega L}\right)^{1/2}\left(1 + \frac{G}{j\omega C}\right)^{-1/2}. \tag{3.32e}$$

3. Distortionless line ($R/L = G/C$)

In distortionless transmission lines, $R/L = G/C$, and the defining equations can be simplified as

$$\gamma = \alpha + j\beta = \sqrt{\frac{C}{L}}(R + j\omega L) \tag{3.33a}$$

$$\alpha = R\sqrt{\frac{C}{L}} \tag{3.33b}$$

$$\beta = \omega\sqrt{LC} \tag{3.33c}$$

$$v_p = \frac{1}{\sqrt{LC}} \tag{3.33d}$$

$$Z_0 = \sqrt{\frac{L}{C}}. \tag{3.33e}$$

3.2.2 TERMINATED LOSSLESS TRANSMISSION LINES

Consider the lossless transmission line shown in Figure 3.5. The voltage and current at any point on the line can be written as

$$V(z) = V_o^+ e^{-j\beta z} + V_o^- e^{j\beta z} \tag{3.34}$$

$$I(z) = \frac{V_o^+}{Z_0} e^{-j\beta z} - \frac{V_o^-}{Z_0} e^{+j\beta z}. \tag{3.35}$$

The voltage and current at the load $z = 0$ in terms of load reflection coefficient is

$$V(z) = V_o^+ \left[e^{-j\beta z} + \Gamma e^{j\beta z} \right] \tag{3.36}$$

$$I(z) = \frac{V_o^+}{Z_0} \left[e^{-j\beta z} - \Gamma e^{j\beta z} \right]. \tag{3.37}$$

It is seen that the voltage and current on the line consist of a superposition of an incident and reflected wave, which represents standing waves. When $\Gamma = 0$, it is a matched condition. The time-average power flow along the line at the point z can be written as

$$P_{avg} = \frac{1}{2}\operatorname{Re}\{V(z)I^*(z)\} = \frac{1}{2}\frac{|V_o^+|^2}{Z_0}\operatorname{Re}\{1 - \Gamma^* e^{-2j\beta z} + \Gamma e^{2j\beta z} - |\Gamma|^2\} \tag{3.38}$$

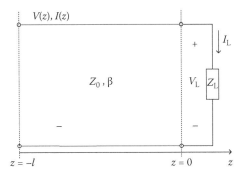

FIGURE 3.5 Lossless transmission line.

or

$$P_{avg} = \frac{1}{2} \frac{\left|V_o^+\right|^2}{Z_0} \left(1 - \left|\Gamma\right|^2\right). \tag{3.39}$$

When the load is mismatched, not all of the available power from the generator is delivered to the load. The power that is lost is known as return loss (RL), and this can be found from

$$RL = -20 \log |\Gamma| \text{ dB}. \tag{3.40}$$

Under mismatched condition, voltage on the line can be written as

$$\left|V(z)\right| = \left|V_o^+\right| \left|1 + \Gamma e^{2j\beta z}\right| = \left|V_o^+\right| \left|1 + \Gamma e^{-2j\beta l}\right| = \left|V_o^+\right| \left|1 + |\Gamma| e^{j(\theta - 2\beta l)}\right|. \tag{3.41}$$

The minimum and maximum values of the voltage from Equation 3.41 is found as

$$V_{max} = \left|V_o^+\right|\left(1 + |\Gamma|\right) \quad \text{and} \quad V_{min} = \left|V_o^+\right|\left(1 - |\Gamma|\right). \tag{3.42}$$

A measure of the mismatch of a line is called the voltage standing wave ratio (VSWR), which can be expressed as the ratio of maximum voltage to minimum voltage as

$$VSWR = \frac{V_{max}}{V_{min}} = \frac{1 + |\Gamma|}{1 - |\Gamma|}. \tag{3.43}$$

From Equation 3.41, the distance between two successive voltage maxima (or minima) is $l = 2\pi/2\beta = \lambda/2$ ($2\beta l = 2\pi$), while the distance between a maximum and a minimum is $l = \pi/2\beta = \lambda/4$. From (3.37) with $z = -l$,

$$\Gamma(l) = \frac{V_o^- e^{-j\beta l}}{V_o^+ e^{j\beta l}} = \Gamma(0)e^{-2j\beta l}. \tag{3.44}$$

For the current,

$$I(z) = V_o^+ e^{-j\beta z} \left(\frac{1}{Z_0}\right)\left(1 - |\Gamma| e^{+j(\phi - 2\beta l)}\right) \tag{3.45a}$$

or

$$\left|I(z)\right| = \left|V_o^+\right| \left(\frac{1}{Z_0}\right)\left|1 - |\Gamma| e^{+j(\phi - 2\beta l)}\right|. \tag{3.45b}$$

Hence, the maximum and minimum values of the current on the line can be written as

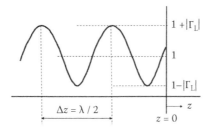

FIGURE 3.6 Voltage versus transmission line length.

$$I_{max} = |I(z)|_{max} = |V_o^+|\left(\frac{1}{Z_0}\right)(1+|\Gamma|) \qquad (3.46a)$$

$$I_{min} = |I(z)|_{min} = |V_o^+|\left(\frac{1}{Z_0}\right)(1-|\Gamma|). \qquad (3.46b)$$

The current standing wave ratio (ISWR) is

$$\text{ISWR} = \frac{I_{max}}{I_{min}} = \frac{1+|\Gamma|}{1-|\Gamma|}. \qquad (3.47)$$

Hence, VSWR = ISWR from Equations 3.43 and 3.47. We will be using VSWR through the book for analysis. The voltage waveform versus length of the transmission line along the axis is plotted in Figure 3.6.

At a distance $l = -z$, the input impedance is the equal to

$$Z_{in} = \frac{V(-l)}{I(-l)} = Z_0\frac{Z_L + jZ_0\tan\beta l}{Z_0 + jZ_L\tan\beta l}. \qquad (3.48)$$

Example 3.1

A 2-m lossless air-spaced transmission line having a characteristic impedance of $30\,\Omega$ is terminated with an impedance of $40 + j30[\Omega]$ at an operating frequency of $200\,\text{MHz}$. Find the input impedance.

Solution: The phase constant is found from

$$\beta = \frac{\omega}{v_p} = \frac{4}{3}\pi.$$

Since it is given that $R_0 = 50\,\Omega$, $Z_L = 40 + j30$, and $\ell = 2m$, we obtain the input impedance from Equation 3.48 as

$$Z_i = 50\frac{(40 + j30) + j50\cdot\tan\left(\frac{4\pi}{3}\cdot 2\right)}{50 + j(40 + j30)\cdot\tan\left(\frac{4\pi}{3}\cdot 2\right)} = 26.3 - j9.87.$$

Example 3.2

For a transmission, it is given that $Z_L = 17.4 - j30\ [\Omega]$ and $Z_0 = 50\ [\Omega]$. Calculate Γ_L, SWR, z_{min}, V_{max}, V_{min} on the transmission line.

Solution: From given information, we find the load reflection coefficient as

$$\Gamma_L = \frac{Z_L - Z_0}{Z_L + Z_0} = -0.24 - j0.55 = 0.6\ e^{-j(1.99)}.$$

Voltage standing wave ratio is found from

$$SWR = \frac{V_{max}}{V_{min}} = \frac{1 + |\Gamma_L|}{1 - |\Gamma_L|} = \frac{1 + 0.6}{1 - 0.6} = 4.0.$$

This leads to

$$V_{max}/|V^+| = 1 + |\Gamma_L| = 1.6$$

$$V_{min}/|V^+| = 1 - |\Gamma_L| = 0.4.$$

Hence, the maximum and minimum value of the voltage are obtained when

$$V_{max}\ \text{when}\ \phi + 2\beta z = 0, -2\pi, ...$$

$$V_{min}\ \text{when}\ \phi + 2\beta z = -\pi, -3\pi, ...$$

So, the distance that will give the minimum value of the voltage is found from

$$z_{min} = \frac{-\pi - \phi}{2\beta} = \frac{(-\pi + 1.99)}{2(2\pi/\lambda)} = -0.092\lambda.$$

When the voltage waveform is plotted versus transmission line length, the results agree with the calculated results as shown in Figure 3.7a.

Example 3.3

SWR on a lossless 30 Ω terminated line terminated in an unknown load impedance is 4. The distance between successive minimum is 30 cm, and the first minimum is located at 6 cm from the load. Determine Γ, Z_L, and l_m.

Solution: From the given information, the wavelength can be found as

$$\frac{\lambda}{2} = 0.3 \Rightarrow \lambda = 0.6\text{m},\ \beta = \frac{2\pi}{\lambda} = 3.33\pi.$$

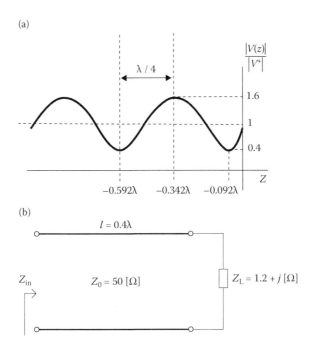

FIGURE 3.7 (a) Voltage versus transmission length, for example and (b) transmission line circuit.

The reflection coefficient is equal to

$$|\Gamma| = \frac{4-1}{4+1} = 0.6, \; z'_m = 0.06 \text{ m} \Rightarrow \ell_m = \frac{\lambda}{2} - z'_m = = 0.24 \text{ m}$$

$$\theta_\Gamma = 2\beta z'_m - \pi = -0.6\pi, \; \Gamma = |\Gamma| e^{j\theta_\Gamma} = 0.6 e^{-j0.6\pi} = -0.185 - j0.95.$$

The load impedance is then equal to

$$Z_L = Z_0 \frac{1+\Gamma_L}{1-\Gamma_L} = 50 \cdot \frac{1+\left(-0.185 - j0.95\right)}{1-\left(-0.185 - j0.95\right)} = 1.43 - j41.17.$$

Example 3.4

Calculate the following parameters for the transmission line shown in Figure 3.7b when a normalized load of $1.2 + j$ [Ω] is connected:

1. The voltage standing wave ratio on the line
2. Load reflection coefficient
3. Admittance of the load
4. Impedance at the input of the line
5. The distance from the load to the first voltage minimum
6. The distance from the load to the first voltage maximum

Solution: Because the load impedance is already normalized, we can skip normalization process and start the calculations as follows:

1. $SWR = 2.5$
2. $\Gamma_L = 0.42\angle 54.5°$
3. $Y_L = \dfrac{y_L}{Z_0} = \dfrac{0.5 - j0.42}{50\ \Omega} = (10 - j8.4)\ \text{mS}$
4. $Z_{in} = z_{in} \cdot Z_0 = (0.5 + j0.4) \cdot Z_0 = (25 + j0.4)\Omega$
5. $l_{min} = 0.5\lambda - 0.174\lambda = 0.326\lambda$
6. $l_{max} = 0.25\lambda - 0.174\lambda = 0.076\lambda$

3.2.3 SPECIAL CASES OF TERMINATED TRANSMISSION LINES

1. *Short-circuited line*: Consider the short-circuited transmission line shown in Figure 3.8. When the transmission line is open-circuited, $Z_L = 0 \rightarrow \Gamma = -1$, then voltage and current can be written as

$$V(z) = V_o^+\left[e^{-j\beta z} - e^{j\beta z}\right] = -2\,jV_o^+ \sin\beta z \qquad (3.49)$$

$$I(z) = \dfrac{V_o^+}{Z_0}\left[e^{-j\beta z} + e^{j\beta z}\right] = 2\dfrac{V_o^+}{Z_0}\cos\beta z. \qquad (3.50)$$

The input impedance when $z = -l$ is the equal to

$$Z_{in} = jZ_0 \tan\beta l. \qquad (3.51)$$

The impedance variation of the line along the z is given in Figure 3.9.
At lower frequencies, Equation 3.51 can be written as

$$X_{in} \approx Z_0\,(\beta l) = \sqrt{\dfrac{L}{C}}\left(\omega\sqrt{LC}\,l\right) = \omega(Ll). \qquad (3.52)$$

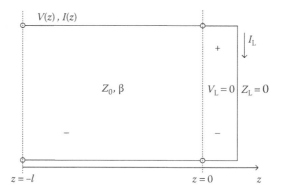

FIGURE 3.8 Short-circuited transmission line.

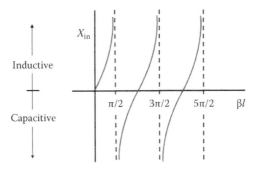

FIGURE 3.9 Impedance variation for a short-circuited transmission line.

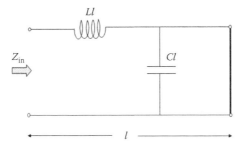

FIGURE 3.10 Low-frequency equivalent circuit of a short-circuited transmission line.

Then, the lumped element equivalent model of the transmission line can be represented as shown in Figure 3.10.

2. *Open-circuited line:* Consider the open-circuited transmission line shown in Figure 3.11. When the transmission line is open-circuited, $Z_L = \infty \rightarrow \Gamma = 1$, then voltage and current can be written as

$$V(z) = V_o^+\left[e^{-j\beta z} + e^{j\beta z}\right] = 2V_o^+ \cos\beta z \tag{3.53}$$

$$I(z) = \frac{V_o^+}{Z_0}\left[e^{-j\beta z} - e^{j\beta z}\right] = \frac{-2jV_o^+}{Z_0}\sin\beta z. \tag{3.54}$$

The input impedance when $z = -l$ is the equal to

$$Z_{in} = -jZ_0 \cot\beta l. \tag{3.55}$$

The impedance variation of the line along the z is given in Figure 3.12.
At lower frequencies, Equation 3.51 can be written as

$$X_{in} \approx -Z_0/(\beta l) = -\sqrt{\frac{L}{C}}\left(\frac{1}{\omega\sqrt{LC}\,l}\right) = \frac{-1}{\omega(Cl)}. \tag{3.56}$$

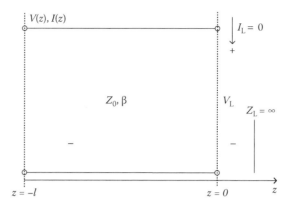

FIGURE 3.11 Open-circuited transmission line.

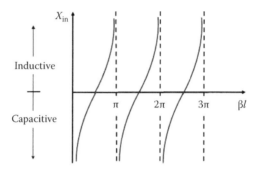

FIGURE 3.12 Impedance variation for an open-circuited transmission line.

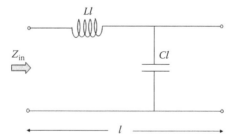

FIGURE 3.13 Low-frequency equivalent circuit of an open-circuited transmission line.

Then, the lumped element equivalent model of the transmission line can be represented as shown in Figure 3.13.

3.3 SMITH CHART

Smith chart is a conformal mapping between the normalized complex impedance plane and the complex reflection coefficient plane. It is a graphical method of displaying impedances and all related parameters using reflection coefficient. It was invented by Phillip Hagar Smith while he was working at Radio Corporation of America. The process of establishing Smith chart begins with normalizing the impedance as shown by Equation 3.57.

$$z_L = \frac{Z_L}{Z_0} = \frac{R_L + jX_L}{Z_0}. \tag{3.57}$$

Consider now the right-hand portion of the normalized complex impedance plane as illustrated in Figure 3.14. All values of impedance such that $R \geq 0$ are represented by points in the plane. The impedance of all passive devices will be represented by points in the right-half plane.

The complex reflection coefficient may be written as a magnitude and a phase or as real and imaginary parts.

$$\Gamma_L = |\Gamma_L| e^{\angle \Gamma_L} = \Gamma_{Lr} + j\Gamma_{Li}. \tag{3.58}$$

The reflection coefficient in terms of the load Z_L terminating line Z_0 is defined as

$$\Gamma_L = \frac{Z_L - Z_0}{Z_L + Z_0}. \tag{3.59}$$

We can rearrange Equation 3.59 to get

$$Z_L = Z_0 \frac{1 + \Gamma_L}{1 - \Gamma_L}. \tag{3.60}$$

In terms of normalized quantities, Equation 3.60 can be written as

$$z_L = r_L + jx_L = \frac{Z_L}{Z_0} = \frac{1 + \Gamma_L}{1 - \Gamma_L}. \tag{3.61}$$

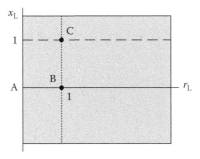

FIGURE 3.14 Right-hand portion of the normalized complex impedance plane.

Substituting in the complex expression for Γ_L and equating real and imaginary parts, we find the two equations, which represent circles in the complex reflection coefficient plane as

$$\left(\Gamma_{Lr} - \frac{r_L}{1+r_L}\right)^2 + \left(\Gamma_{Li} - 0\right)^2 = \left(\frac{1}{1+r_L}\right)^2 \qquad (3.62)$$

$$\left(\Gamma_{Lr} - 1\right)^2 + \left(\Gamma_{Li} - \frac{1}{x_L}\right)^2 = \left(\frac{1}{x_L}\right)^2. \qquad (3.63)$$

The first circle is centered at

$$\left(\frac{r_L}{1+r_L}, 0\right) \qquad (3.64)$$

and the second circle is centered at

$$\left(1, \frac{1}{x_L}\right). \qquad (3.65)$$

The location of the first circle is always inside the unit circle in the complex reflection coefficient plane with corresponding radius as

$$\frac{1}{1+r_L}. \qquad (3.66)$$

Therefore, this circle will always be fully contained within the unit circle because the radius can never be greater than unity. This conformal mapping represents the mapping of the real resistance circle (shown in Figure 3.15) using the following mapping equation:

$$\left(\Gamma_r - \frac{r}{1+r}\right)^2 + \left(\Gamma_L\right)^2 = \left(\frac{1}{1+r}\right)^2. \qquad (3.67)$$

The location of the second circle is always outside the unit circle in the complex reflection coefficient plane with the corresponding radius as

$$\left(\frac{1}{x_L}\right). \qquad (3.68)$$

The value radius can vary between 0 and ∞. This conformal mapping represents the mapping of the imaginary reactance circle and is shown in Figure 3.16, using the following mapping equation:

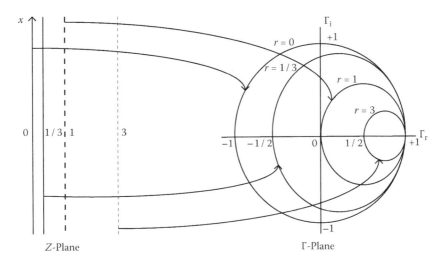

FIGURE 3.15 Conformal mapping of constant resistances.

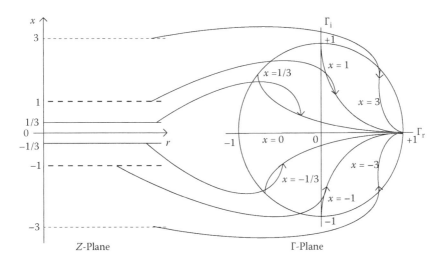

FIGURE 3.16 Conformal mapping of constant reactances.

$$\left(\Gamma_r - 1\right)^2 + \left(\Gamma_i - \frac{1}{x}\right)^2 = \left(\frac{1}{x}\right)^2. \tag{3.69}$$

The circles centered on the real axis represent lines of constant real part of the load impedance (r_L is constant, x_L varies), and the circles whose centers reside outside the unit circle represent lines of constant imaginary part of the load impedance (x_L is constant, r_L varies). Combining the results of two mapping into a single mapping gives the display of complete Smith chart as shown in Figure 3.17.

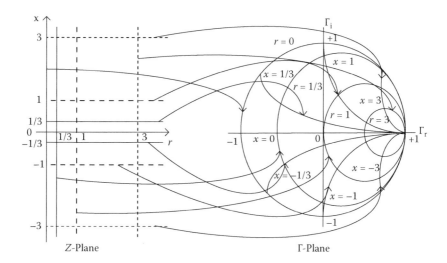

FIGURE 3.17 Combined conformal mapping lead to the display of Smith chart.

In summary, the properties of the r-circles are as follows:

- The centers of all r-circles lie on the Γ_r-axis.
- The $r = 0$ circle, having a unity radius and centered at the origin, is the largest.
- The r-circles become progressively smaller as r increases from 0 toward ∞, ending at the $(\Gamma_r = 1, \Gamma_i = 0)$ point for open circuit.
- All r-circles pass through the $(\Gamma_r = 1, \Gamma_i = 0)$ point.

Similarly, the properties of the x-circles are as follows:

- The centers of all x-circles lie on the $\Gamma_r = 1$ line, those for $x > 0$ (inductive reactance) lie above the Γ_r-axis, and those for $x < 0$ (capacitive reactance) lie below the Γ_r-axis.
- The $x = 0$ circle becomes the Γ_r-axis.
- The x-circle becomes progressively smaller as $|x|$ increases from 0 toward ∞, ending at the $(\Gamma_r = 1, \Gamma_i = 0)$ point for open circuit.
- All x-circles pass through the $(\Gamma_r = 1, \Gamma_i = 0)$ point.

Hence, in the combined display of the Smith chart

- All $|\Gamma|$ circles are centered at the origin, and their radii vary uniformly from 0 to 1.
- The angle, measured from the positive real axis, of the line drawn from the origin through the point representing z_L equals $\theta\Gamma$.
- The value of the r-circle passing through the intersection of the $|\Gamma|$ circle and the positive real axis equals the SWR.

Example 3.5

Locate the following normalized impedances on the Smith chart, and calculate standing wave ratios and reflection coefficients (1) $z = 0.2 + j0.3$, (2) $z = 0.4 + j0.7$, and (3) $z = 0.6 + j0.1$

Solution: The following generic MATLAB® code is developed to calculate mark impedance points, draw VSWR circles, and calculate reflection coefficients on the Smith chart at single frequency.

```
%This program marks impedance points, draws VSWR circle,
calculates
%reflection coefficients, and marks them on the Smith Chart at
single
%frequency

clear all;
close all;
global Z0;
Set_Z0(1); %Set Z0 to 1

%Gui Prompt
 prompt = {'ZL1', 'ZL2: ','ZL3'};
dlg_title = 'Enter Impedance ';
num_lines = 1;
def = {'0.2+j*0.3','0.4+j*0.7','0.6+j*0.1'};
answer = inputdlg(prompt,dlg_title,num_lines,def, 'on');
%convert the strings received from the GUI to numbers
valuearray=str2double(answer);

%Give variable names to the received numbers
ZL1=valuearray(1);
ZL2=valuearray(2);
ZL3=valuearray(3);

%part a
gamma1=(ZL1-Z0)/(ZL1+Z0);
VSWR1=(1+abs(gamma1))/(1-abs(gamma1));
[th1,rl1]=cart2pol(real(gamma1),imag(gamma1));
smith; %Call Smith Chart Program
s_point(ZL1);
const_SWR_circle(ZL1,'r--');
hold on;
text(real(gamma1)+0.04,imag(gamma1)-0.03,'\bf\Gamma_1');

%part b
gamma2=(ZL2-Z0)/(ZL2+Z0);
VSWR2=(1+abs(gamma2))/(1-abs(gamma2));
[th2,rl2]=cart2pol(real(gamma2),imag(gamma2));
s_point(ZL2);

%part c
gamma3=(ZL3-Z0)/(ZL3+Z0);
VSWR3=(1+abs(gamma3))/(1-abs(gamma3));
[th3,rl3]=cart2pol(real(gamma3),imag(gamma3));
s_point(ZL3);
```

```
const_SWR_circle(ZL3,'r--');
hold on;
text(real(gamma3)+0.04,imag(gamma3)-0.03,'\bf\Gamma_3');

msgbox( sprintf([...
        'Calculated Parameters for Z1 \n'...
        '   Reflection coefficient for Z1:  gamma1 =%f
+j(%f)\n'...
        '   Reflection Coefficent for Z1 In Polar form
:|gamma1|=%f,angle1=%f\n'...
        '   Standing Wave Ratio for Z1 : VSWR1=%f \n'...
        '\n'...
        'Calculated Parameters for Z2 \n'...
        '   Reflection coefficient for Z2:  gamma2 =%f
+j(%f)\n'...
        '   Reflection Coefficient for Z2 In Polar form
:|gamma2|=%f,angle1=%f\n'...
        '   Standing Wave Ratio for Z2 : VSWR2=%f \n'...
        '\n'...
        'Calculated Parameters for Z3 \n'...
        '   Reflection coefficient for Z3:  gamma3 =%f
+j(%f)\n'...
        '   Reflection Coefficient for Z3 In Polar form
:|gamma3|=%f,angle3=%f\n'...
        '   Standing Wave Ratio for Z3 : VSWR3=%f \n'...
          '\n']...
,real(gamma1),imag(gamma1),rl1,th1*180/pi,VSWR1,real(gamma2),imag
(gamma2),rl2,th2*180/pi,VSWR2,real(gamma3),imag(gamma3),rl3
,th3*180/pi,VSWR3));
```

When the program is executed, the GUI is displayed as shown in Figure 3.18a for entering impedances and result.

The results are displayed on Smith chart in Figure 3.18b.

3.3.1 Input Impedance Determination with Smith Chart

It was shown earlier that the voltage and current at any point on the transmission line can be expressed as

$$V(z') = \frac{I_L}{2}(Z_L + Z_0)e^{\gamma z'}\left[1 + \Gamma e^{-2\gamma z'}\right] \tag{3.70}$$

$$I(z') = \frac{I_L}{2Z_0}(Z_L + Z_0)e^{\gamma z'}\left[1 - \Gamma e^{-2\gamma z'}\right], \tag{3.71}$$

where $z' = 1 - z$. Then, input impedance at a distance d away from the load on the line in terms of reflection coefficient can be obtained as

$$Z(d) = \frac{V(d)}{I(d)} = Z_0\frac{1 + \Gamma(d)}{1 - \Gamma(d)}, \tag{3.72}$$

where

$$\Gamma(d) = \Gamma_L e^{-j2\beta d}. \tag{3.73}$$

Example 3.6

A transmission line of characteristic impedance $Z_0 = 30\ \Omega$ and length $d = 0.2\lambda$ is terminated into a load impedance of $Z_L = (23 - j30)\Omega$. Find Γ_L, Z_{in} (d), and SWR using Smith chart.

Solution: The following generic MATLAB code is developed to find input impedance by moving toward generator at any length:

```
%This program find input impedance by moving towards generator
%on the transmission line at single frequency at any length
```

(a)

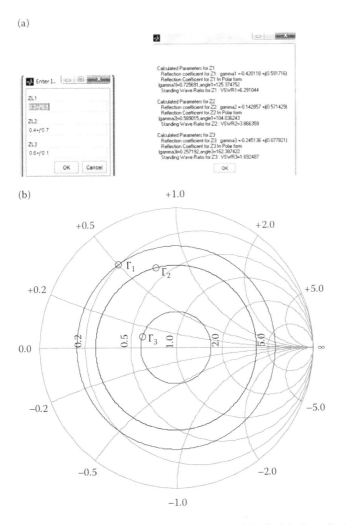

(b)

FIGURE 3.18 (a) GUI display for user input and results and (b) Smith chart displaying the calculated values and impedances.

```
clear all;
close all;
global Z0;
%Gui Prompt

prompt = {'Ente Load Impedance ZL :', 'Enter the Length (in
lambda) d : ','Enter Characteristic Impedance Z0:'};
dlg_title = 'Enter Impedance ';
num_lines = 1;
def = {'23-j*30','.2','30'};
answer = inputdlg(prompt,dlg_title,num_lines,def, 'on');
%convert the strings received from the GUI to numbers
valuearray=str2double(answer);

%Give variable names to the received numbers
ZL=valuearray(1);
d=valuearray(2);
Z0=valuearray(3);


Set_Z0(Z0);
gamma_0=(ZL-Z0)/(ZL+Z0);
[th0,mag_gamma_0]=cart2pol(real(gamma_0),imag(gamma_0));
if th0<0
    th0=th0+2*pi;
end
th_in=th0-2*2*pi*d;
if th_in<0
    th_in=th_in+2*pi;
end



[x_gamma_in,y_gamma_in]=pol2cart(th_in,mag_gamma_0);
Zin=Z0*(1+x_gamma_in+j*y_gamma_in)/(1-x_gamma_in-j*y_gamma_in);
SWR=(1+abs(gamma_0))/(1-abs(gamma_0));
smith_chart(0);
hold on;
th=th0:(th_in-th0)/29:th_in;
gamma=mag_gamma_0*ones(1,30);
polar(th,gamma,'k');
hold on
s_point(Zin);
text(x_gamma_in+0.04,y_gamma_in-0.03,'\bfZ_{in}');
s_point(ZL);
text(real(gamma_0)+0.04,imag(gamma_0)-0.03,'\bfZ_{L}');

msgbox( sprintf([...
        'Calculated Parameters for Transmission Line \n'...
        '   Load Reflection coefficient :   gamma_0 =%f
+j(%f)\n'...
        '   Magnitude of Load Reflection Coefficient
:|gamma_0|=%f,angle=%f\n'...
        '    Input Impedance Zin : Zin=%f +j(%f)\n'...
        '    Standing Wave Ratio : SWR=%f \n'...
        '\n']...
```

```
,real(gamma_0),imag(gamma_0),mag_gamma_0,th0*180/
pi,real(Zin),imag(Zin),SWR));
```

When the program is executed, the following GUI and calculated results are displayed. The program also displays input impedance on the Smith chart as shown in Figure 3.19b with the move toward the generator.

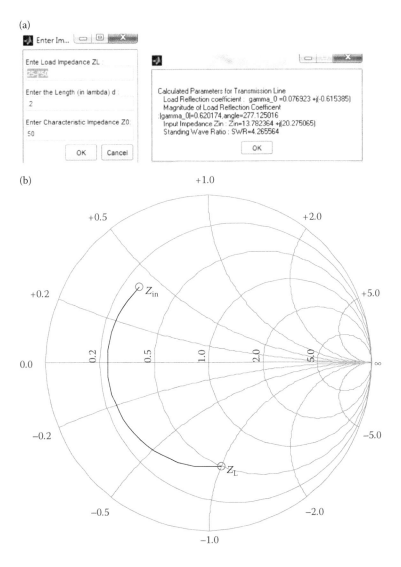

FIGURE 3.19 (a) MATLAB GUI display for user input and results and (b) input impedance display using Smith chart.

3.3.2 Smith Chart as an Admittance Chart

Smith chart can be also used as an admittance chart by transforming impedances to admittances. Consider the expression for a normalized impedance at any point on the transmission line in terms of reflection coefficient as

$$Z_{in}(z) = \left(\frac{1 + \Gamma(z)}{1 - \Gamma(z)} \right). \tag{3.74}$$

The normalized admittance is the reciprocal of impedance and can be written as

$$Y_{in}(z) = \frac{Y_{in}(z)}{Y_0} = \frac{1/Z_{in}(z)}{1/Z_0} = \frac{1}{Z_{in}(z)/Z_0} = \frac{1}{Z_{in}(z)}. \tag{3.75}$$

Then, the normalized admittance in terms of reflection coefficient can be expressed as

$$Y_{in}(z) = \left(\frac{1 - \Gamma(z)}{1 + \Gamma(z)} \right), \tag{3.76}$$

which can be written as

$$Y_{in}(z) = \left(\frac{1 + \Gamma'(z)}{1 - \Gamma'(z)} \right), \tag{3.77}$$

where

$$\Gamma'(z) = -\Gamma(z) = \Gamma(z)e^{-j\pi}. \tag{3.78}$$

That means a 180° phase shift for the reflection coefficient gives the value of admittance for the corresponding impedance value. When the impedance point is marked on the Smith chart, moving 180° clockwise gives the value admittance. Instead of repeating this for each impedance point on the Smith chart, we can keep the location of the impedance fixed and rotate the Smith chart by 180°. This gives the admittance chart as shown in Figure 3.20. When both Z and Y chart are plotted together, we obtain ZY chart as shown in Figure 3.21.

3.3.3 ZY Smith Chart and Its Application

ZY Smith chart gives the ability to implement both impedances and admittances on a single chart. It is a powerful chart, and it helps the designer to make impedance transformation and matching using a unique graphical display when components are connected in series or in shunt. The effect of adding single reactive component in series with a complex impedance results in motion along a constant resistance circle in ZY chart. If a single reactive component is added with a complex impedance

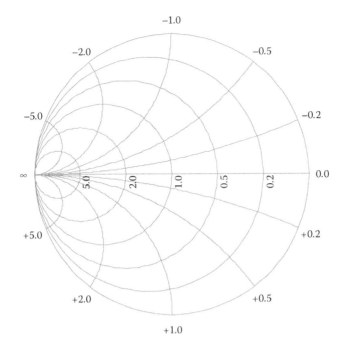

FIGURE 3.20 Admittance, *Y*, Smith chart.

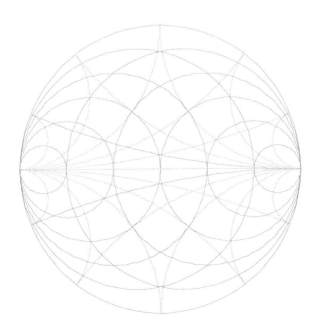

FIGURE 3.21 *ZY* Smith chart.

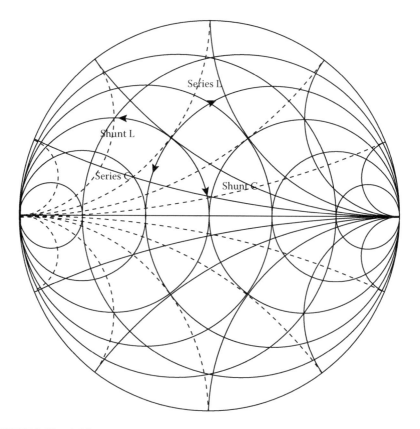

FIGURE 3.22   Adding component using ZY Smith chart.

in shunt, then motion along a constant conductance circle in ZY chart is needed. Whenever an inductor is connected to the network, the direction of movement on ZY chart is towards the upper half, whereas a capacitive involvement results in a movement towards the lower part of the chart. All these component motions on ZY chart are illustrated in Figure 3.22.

Example 3.7

Using the Smith chart, find the input impedance for the circuit shown in Figure 3.23 at 4 GHz when the load connected is $Z_L = 62.3\ \Omega$.

Solution: The process begins with normalizing the load impedance, $Z_L = R = 62.3\ \Omega$:

$$z_L = \frac{Z_L}{Z_0} = \frac{62.5}{50} = 1.25.$$

Because the next component is a shunt-connected component, we need to convert this value to a conductance value. That is

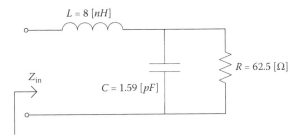

FIGURE 3.23 Impedance transformation.

$$g_L = \frac{1}{z_L} = \frac{1}{1.25} = 0.8.$$

On ZY Smith chart, we mark this point on the conductance circle. The next component is shunt C with a value of 1.39 [pF]. The normalized susceptance value of the capacitor at 4 GHz is found from

$$b_C = B_C Z_0 = \omega C Z_0 = (2\pi 4 \times 10^9)(1.59 \times 10^{-12}) 50 = 2.$$

This corresponds to point B on Smith chart. This is the amount of rotation that needs to be done on the conductance circle as shown by point B. The admittance at point B is equal to

$$y_B = 0.8 + j2.$$

The next component connected is a series L with a value of 8 [nH]. So, we move from conductance circle to resistance circle and read the corresponding impedance value as

$$z_B = 0.17 - j0.43.$$

The normalized reactance value of inductor is equal to

$$x_L = \frac{X_L}{Z_0} = \frac{(2\pi 4 \times 10^9)(8 \times 10^{-9})}{50} = 4.$$

This value needs to be added to the impedance at Point B to find the impedance value shown as Point C on Smith chart.

$$z_C = z_B + x_L = 0.17 - j0.43 + 4 = 0.17 + 3.57.$$

Denormalizing impedance z_C gives the input impedance as

$$Z_{in} = z_C Z_0 = (0.17 + j3.57) 50 = (8.5 + j178.5)[\Omega].$$

The results are shown on Smith chart in Figure 3.24.

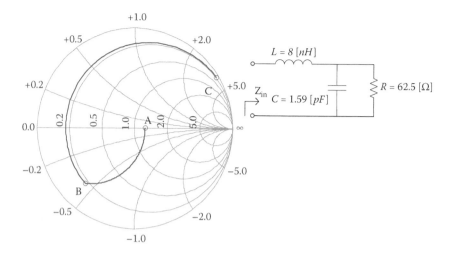

FIGURE 3.24   Impedance transformation using Smith chart.

3.4   IMPEDANCE MATCHING BETWEEN TRANSMISSION LINES AND LOAD IMPEDANCES

Consider the matching network between load and transmission line shown in Figure 3.25. We can implement the matching network using the lumped element L-type sections consisting of two reactive elements. There are eight possible L matching networks that are shown in Figure 3.26. These can be illustrated by generic two circuits as shown in Figure 3.27.

In both the configurations of Figure 3.27, the reactive elements may be either inductors or capacitors. As a result, there are eight distinct possibilities as shown in Figure 3.26 for the matching circuit for various load impedances. If the normalized load impedance, $z_L = Z_L/Z_0$, is inside the $1 + jx$ circle on the Smith chart, then the circuit of Figure 3.27a should be used. If the normalized load impedance is outside the $1 + jx$ circle on the Smith chart, the circuit of Figure 3.27b should be used. The $1 + jx$ circle is the resistance circle on the impedance Smith chart for which $r = 1$.

Consider first the circuit given in Figure 3.27a with $Z_L = R_L + jX_L$. It is assumed that $R_L > Z_0$ and $z_L = Z_L/Z_0$ maps inside the $1 + jx$ circle on the Smith chart. For a

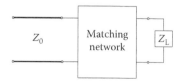

FIGURE 3.25   Matching network between load and transmission line.

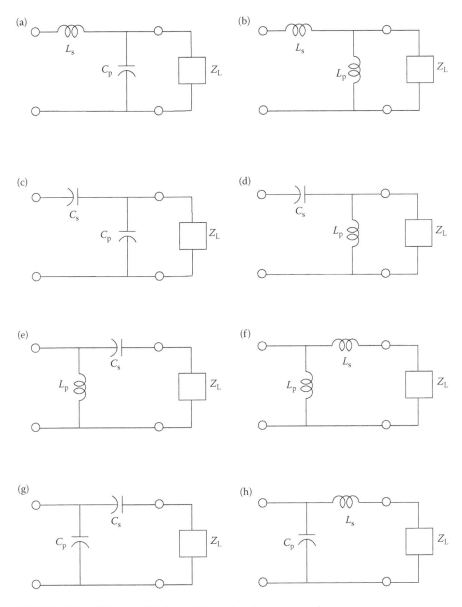

FIGURE 3.26 Eight possible L matching network sections.

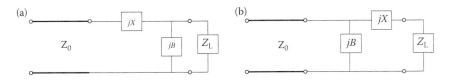

FIGURE 3.27 Generic L matching network sections to represent eight L sections.

matched condition, the impedance seen looking into the matching network followed by the load impedance is then equal to Z_0 and can be written as

$$Z_0 = jX + \frac{1}{jB + \left(\dfrac{1}{R_L} + jX_L\right)}. \tag{3.79}$$

Separating Equation 3.79 into real and imaginary parts gives two equations with two unknowns, X and B, as

$$B\left(XR_L - X_L Z_0\right) = R_L - Z_0 \tag{3.80}$$

$$X\left(1 - BX_L\right) = BZ_0 R_L - X_L. \tag{3.81}$$

Solution of Equations 3.80 and 3.81 lead to

$$B = \frac{X_L \pm \sqrt{\dfrac{R_L}{Z_0}} \sqrt{R_L^2 + X_L^2 - Z_0 R_L}}{R_L^2 + X_L^2} \tag{3.82}$$

and

$$X = \frac{1}{B} + \frac{X_L Z_0}{R_L} - \frac{Z_0}{BR_L}. \tag{3.83}$$

From Equation 3.82, there exist two possibilities for B and consequently for X. Both of these solutions are physically realizable, and constitute all the values of B and X. Positive value of X gives an inductor, negative value of X gives a capacitor. Similarly, positive value of B gives a capacitor and negative value of B gives an inductor.

We can repeat the same procedure for the generic L matching network shown in Figure 3.27b. This circuit is used when $z_L = Z_L/Z_0$ and it maps outside the $1 + jx$ circle on the Smith chart since it is assumed that $R_L < Z_0$. For a matched condition, the admittance seen looking into the matching network followed by the load impedance $Z_L = R_L + jX_L$ is then equal to $1/Z_0$ and can be written as

$$\frac{1}{Z_0} = jB + \frac{1}{R_L + j(X + X_L)}. \tag{3.84}$$

Separating Equation 3.84 into real and imaginary parts gives the following two equations with two unknowns, X and B, as

$$BZ_0\left(X + X_L\right) = Z_0 - R_L \tag{3.85}$$

$$\left(X + X_L\right) = BZ_0 R_L. \tag{3.86}$$

Solution of Equations 3.85 and 3.86 lead to

$$X = \sqrt{R_L(Z_0 - R_L)} - X_L \tag{3.87}$$

$$B = \pm \frac{\sqrt{(Z_0 - R_L)/R_L}}{Z_0}. \tag{3.88}$$

Equation 3.88 has two possible solutions for B.

In order to match an arbitrary complex load to a line of characteristic impedance Z_0, the real part of the input impedance to the matching network must be Z_0, whereas the imaginary part must be zero. This implies that a general matching network must have at least two degrees of freedom; in the L section matching circuit, these two degrees of freedom are provided by the values of the two reactive components.

3.5 SINGLE STUB TUNING

At high frequencies, it may be desirable to match the given load to the transmission line using transmission lines instead of lumped-element components discussed in the previous section. Impedance matching can then be done using a single open-circuited or short-circuited length of transmission line called "stub." It is connected either in parallel or in series with the transmission feed line at a certain distance from the load as shown in Figure 3.28.

In single-stub tuning, there are two design parameters: the distance, d, from the load to the stub position, and the value of susceptance or reactance provided by the shunt or series stub.

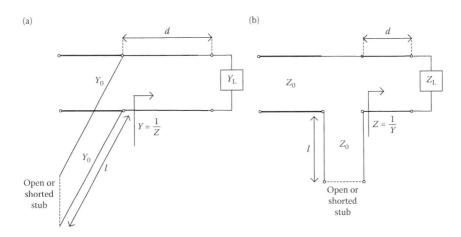

FIGURE 3.28   Single stub matching (a) parallel (b) series.

3.5.1 SHUNT SINGLE STUB TUNING

When it is a shunt-stub case as shown in Figure 3.28a, we select d so that the admittance, Y, seen looking into the line at distance, d, from the load is equal to $Y_0 + jB$. Then, the matching is done by choosing the stub susceptance as $-jB$.

To obtain the relations for d and l, we write the input impedance, $Z_L = 1/Y_L = R_L + jX_L$ at a distance d from the load as

$$Z = Z_0 \frac{(R_L + jX_L) + jZ_0 \tan \beta d}{Z_0 + j(R_L + jX_L)\tan \beta d}. \tag{3.89}$$

We then write the admittance from Equation 3.89 as

$$Y = G + jB = \frac{1}{Z}, \tag{3.90}$$

where

$$G = \frac{R_L(1 + \tan^2 \beta d)}{R_L^2 + (X_L + Z_0 \tan \beta d)^2} \tag{3.91}$$

$$B = \frac{R_L^2 \tan \beta d - (Z_0 - X_L \tan \beta d)(X_L + Z_0 \tan \beta d)}{Z_0[R_L^2 + (X_L + Z_0 \tan \beta d)^2]}. \tag{3.92}$$

To have the matching condition, we need to set G in Equation 3.91 to $G = Y_0 = 1/Z_0$. Hence,

$$Z_0(R_L - Z_0)\tan^2 \beta d - 2X_L Z_0 \tan \beta d + (R_L Z_0 - R_L^2 - X_L^2) = 0, \tag{3.93}$$

which lead to two solutions for $\tan \beta d$ as

$$\tan \beta d = \frac{X_L \pm \sqrt{R_L[(Z_0 - R_L)^2 + X_L^2]/Z_0}}{R_L - Z_0}, \quad \text{for } R_L \neq Z_0. \tag{3.94}$$

If $R_L = Z_0$, then $\tan\beta d = -X_L/2Z_0$. As a result, we have solutions for d as

$$\frac{d}{\lambda} = \begin{cases} \dfrac{1}{2\pi}\tan^{-1}\left(-\dfrac{X_L}{2Z_0}\right) & \text{for } -\dfrac{X_L}{2Z_0} \geq 0 \\[3mm] \dfrac{1}{2\pi}\left(\pi + \tan^{-1}\left(-\dfrac{X_L}{2Z_0}\right)\right) & \text{for } -\dfrac{X_L}{2Z_0} < 0 \end{cases}. \tag{3.95}$$

To find the required stub lengths, we first set $B_s = -B$. This leads to the final solutions for open and shorted stubs shown in Figure 3.28a as

$$\frac{l}{\lambda} = \frac{1}{2\pi} \tan^{-1}\left(\frac{B_s}{Y_0}\right) = -\frac{1}{2\pi} \tan^{-1}\left(\frac{B}{Y_0}\right) \text{ for open stub} \tag{3.96}$$

$$\frac{l}{\lambda} = -\frac{1}{2\pi} \tan^{-1}\left(\frac{Y_0}{B_s}\right) = \frac{1}{2\pi} \tan^{-1}\left(\frac{Y_0}{B}\right) \text{ for shorted stub.} \tag{3.97}$$

Smith chart solution for the matching with open stub is practical and can be described as follows:

- Normalize the load impedance, and locate corresponding admittance on the Z smith chart.
- Rotate clockwise around the Smith chart from y_L until it intersects the $g = 1$ circle. It intersects $g = 1$ circle at two points. The "length" of this rotation determines the value d. There are two possible solutions.
- Rotate clockwise from the short/open circuit point around the $g = 0$ circle, until stub b equals $-b$. The "length" of this rotation determines the stub length l.

3.5.2 SERIES SINGLE STUB TUNING

For the series stub case shown in Figure 3.28b, d is chosen so that the impedance looking into the line at a distance d from the load is equal to $Z_0 + jX$. Then, we select the stub reactance is to be $-j X$ to match the line.

To obtain the relations for d and l, we write the input impedance and the input admittance, $Y_L = 1/Z_L = G_L + jB_L$, at a distance d from the load as

$$Y = Y_0 \frac{(G_L + jB_L) + jY_0 \tan\beta d}{Y_0 + j(G_L + jB_L)\tan\beta d}. \tag{3.98}$$

We then write the impedance from Equation 3.98 as

$$Z = R + jX = \frac{1}{Y}, \tag{3.99}$$

where

$$R = \frac{G_L(1 + \tan^2\beta d)}{G_L^2 + (B_L + Y_0 \tan\beta d)^2} \tag{3.100}$$

$$X = \frac{G_L^2 \tan\beta d - (Y_0 - B_L \tan\beta d)(B_L + Y_0 \tan\beta d)}{Y_0[G_L^2 + (B_L + Y_0 \tan\beta d)^2]}. \tag{3.101}$$

To have the matching condition, we need to set G in Equation 3.100 to $R = Z_0 = 1/Y_0$. Therefore,

$$Y_0(G_L - Y_0)\tan^2\beta d - 2B_L Y_0 \tan\beta d + (G_L Y_0 - G_L^2 - B_L^2) = 0, \tag{3.102}$$

which leads to two solutions for $\tan\beta d$ as

$$\tan\beta d = \frac{B_L \pm \sqrt{G_L[(Y_0 - G_L)^2 + B_L^2]/Y_0}}{G_L - Y_0}, \quad \text{for } G_L \neq Y_0. \tag{3.103}$$

If $G_L = Y_0$, then $\tan\beta d = -B_L/2Y_0$. As a result, we have solutions for d as

$$\frac{d}{\lambda} = \begin{cases} \dfrac{1}{2\pi}\tan^{-1}\left(-\dfrac{B_L}{2Y_0}\right) & \text{for } -\dfrac{B_L}{2Y_0} \geq 0 \\[4mm] \dfrac{1}{2\pi}\left(\pi + \tan^{-1}\left(-\dfrac{B_L}{2Y_0}\right)\right) & \text{for } -\dfrac{B_L}{2Y_0} < 0 \end{cases}. \tag{3.104}$$

To find the required stub lengths, we first set $X_s = -X$. This leads to the final solutions for open and shorted stubs shown in Figure 3.28b as

$$\frac{l}{\lambda} = \frac{1}{2\pi}\tan^{-1}\left(\frac{X_S}{Z_0}\right) = -\frac{1}{2\pi}\tan^{-1}\left(\frac{X}{Z_0}\right) \tag{3.105}$$

$$\frac{l}{\lambda} = -\frac{1}{2\pi}\tan^{-1}\left(\frac{Z_0}{X_S}\right) = \frac{1}{2\pi}\tan^{-1}\left(\frac{Z_0}{X}\right). \tag{3.106}$$

Smith chart solution for the matching with series stub is practical and can be described as follows:

- Normalize load impedance, and locate it on the Z Smith chart.
- Rotate clockwise around the Smith chart from z_L until it intersects the $r = 1$ circle. It intersects $r = 1$ circle at two points. The "length" of this rotation determines the value d. There are two possible solutions.
- Rotate clockwise from the short/open circuit point around the $r = 0$ circle, until stub x equals $-x$. The "length" of this rotation determines the stub length l.

3.6   IMPEDANCE TRANSFORMATION AND MATCHING BETWEEN SOURCE AND LOAD IMPEDANCES

Consider the matching network between source and load as shown in Figure 3.29. As discussed in Section 3.4, there are eight possible matching networks as shown in Figure 3.26, which can be represented by generic two L-type matching networks as shown in Figure 3.30. We will first derive the analytical equations as we did in the previous section. This time, consider first the generic L-type matching network shown in Figure 3.30b.

Because source is matched to load impedance, complex conjugate impedance of the load should be equal to the overall impedance connected to the load impedance. This can be expressed by

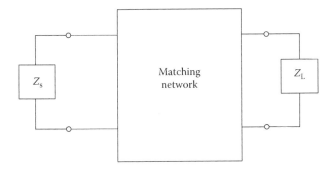

FIGURE 3.29 Matching network between load and transmission line.

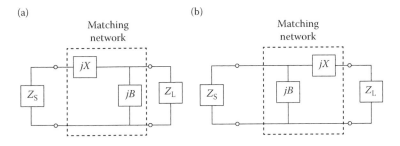

FIGURE 3.30 Generic two L matching networks between source and load impedances.

$$Z_L^* = \frac{1}{Z_s^{-1} + jB} + jX \qquad (3.107)$$

Express

$$Z_s = R_s + jX_s \quad \text{and} \quad Z_L = R_L + jX_L \qquad (3.108)$$

Then,

$$Z_{\text{Load}}^* = \frac{1}{Z_s^{-1} + jB} + jX = \frac{R_s + jX_s}{1 + jB(R_s + jX_s)} + jX = R_L - jX_L. \qquad (3.109)$$

Separate real and imaginary parts

$$R_s = R_L(1 - BX_s) + (X_L + X)BR_s \qquad (3.110)$$

$$X_s = R_s R_L B - (1 - B_C X_L)(X_L + X). \qquad (3.111)$$

Solving for B and X gives

$$X = -X_L \pm \sqrt{R_L (R_s - R_L) + \frac{R_L}{R_s} X_s^2}$$ (3.112a)

$$B = \frac{R_s - R_L}{R_s X + R_s X_L - R_L X_s}.$$ (3.112b)

Solution given by Equation 3.112 is valid only when $R_s > R_L$. Similar procedure can be applied for the circuit shown in Figure 3.30a. We obtain the following equations for reactance and susceptance by assuming $R_s < R_L$:

$$B = \frac{R_s X_L \pm \sqrt{R_s R_L \left(R_L^2 + X_L^2 - R_s R_L \right)}}{R_s \left(R_L^2 + X_L^2 \right)}$$ (3.113a)

$$X = \frac{\left(R_L^2 + X_L^2 \right) B - X_L + \dfrac{X_s}{R_s} R_L}{\left(R_L^2 + X_L^2 \right) B^2 - 2 X_L B + 1}.$$ (3.113b)

As it is seen, the analytical calculation of the impedance transformation and matching is tedious. Instead, we can use Smith chart for the same task. For this, there is a standard procedure that needs to be followed. The design procedure for matching source impedance to a load impedance using Smith chart instead of analytical method can be outlined as follows:

- Normalize the given source and complex conjugate load impedances and locate them on the Smith chart.
- Plot constant resistance and conductance circles for the impedances located.
- Identify the intersection points between the constant resistance and conductance circle for impedances located.
- The number of the intersection point corresponds to the number of possible L matching networks.
- By following the paths that go through intersection points, calculate normalized reactances and susceptances.
- Calculate the actual values of the inductors and capacitors by denormalizing at the given frequency.

Example 3.8

Using Smith chart, design all possible configurations of two-element matching networks that match source impedance $Z_s = (13 + j30)\Omega$ to the load $Z_L = (20 - j30)\,\Omega$. Assume characteristic impedance of $Z_0 = 30\,\Omega$ and an operating frequency of $f = 4\,\text{GHz}$.

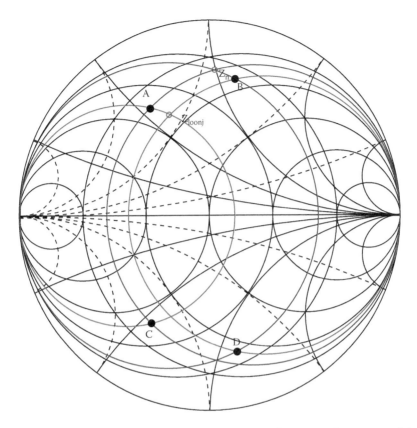

FIGURE 3.31 Number of possible L matching networks to match source and load impedance.

Solution: MATLAB program developed previously is modified to plot resistance and conductance circles for then source and complex conjugate of the load impedances. As shown in Figure 3.31, there are four possible L matching networks. These networks are illustrated in Figure 3.32.

3.7   RESONATOR NETWORKS

Resonators have frequency characteristics that give them the ability to present specific impedance, quality factor, and bandwidth. They can eliminate the reactive component effects and introduce only resistive portion of the impedance at a frequency called resonance frequency. A circuit that is capable of producing these effects is called a resonant circuit. An ideal resonant circuit acts like a filter and eliminates the unwanted signal content out of the frequency of interest as shown in Figure 3.33. Resonant circuits can also be used as part of the impedance matching networks to transform one impedance at one point to another impedance. In RF amplifier circuits, it is a commonly used technique to present a matched impedance at one frequency and introduce high impedance levels at others. When the amplifier is matched at the input and output for maximum gain, it is possible to deliver the

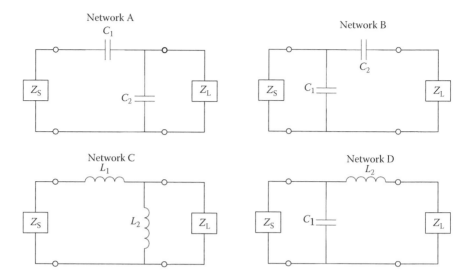

FIGURE 3.32 Possible L matching networks to match source and load impedance.

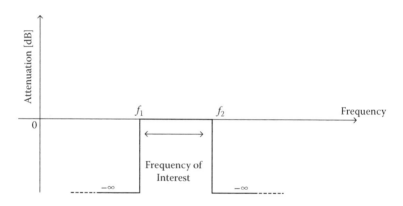

FIGURE 3.33 Ideal resonant network response.

highest amount of power by keeping the circuit stable. The stability of the circuit is mostly accomplished by using filters and resonators to eliminate the spurious contents and oscillations.

In this chapter, resonant networks, transmission lines, Smith chart, and impedance matching networks are discussed.

3.7.1 PARALLEL AND SERIES RESONANT NETWORKS

3.7.1.1 Parallel Resonance

Consider the parallel resonant circuit given in Figure 3.34. The response of the circuit can be obtained by finding the voltage with an application of KCL. Application of KCL gives the first-order differential equation for voltage v as

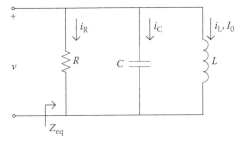

FIGURE 3.34 Parallel resonant circuit.

$$\frac{v}{R}+\frac{1}{L}\int_0^t v\,d\tau+I_o+C\frac{dv}{dt}=0,\qquad(3.114)$$

where I_o is the initial charged current on inductor. Equation 3.114 can be written as

$$\frac{d^2v}{dt^2}+\frac{1}{RC}\frac{dv}{dt}+\frac{v}{LC}=0.\qquad(3.115)$$

The solution for the voltage in Equation 3.115 will be in the following form:

$$v=Ae^{st},\qquad(3.116)$$

where A is constant and $s=j\omega$. Substituting Equation 3.116 into 3.115 gives

$$Ae^{st}\left(s^2+\frac{s}{RC}+\frac{1}{LC}\right)=0,\qquad(3.117)$$

which can be simplified to

$$s^2+\frac{s}{RC}+\frac{1}{LC}=0.\qquad(3.118)$$

Equation 3.118 is called as the characteristic equation. The roots of the equation are

$$s_1=-\frac{1}{2RC}+\sqrt{\left(\frac{1}{2RC}\right)^2-\left(\frac{1}{LC}\right)}\qquad(3.119)$$

$$s_2=-\frac{1}{2RC}-\sqrt{\left(\frac{1}{2RC}\right)^2-\left(\frac{1}{LC}\right)}.\qquad(3.120)$$

The complete solution for the voltage v is then obtained as

$$v = v_1 + v_2 = A_1 e^{s_1 t} + A_2 e^{s_2 t}. \tag{3.121}$$

We can express the roots given by Equations 3.119 and 3.120 as

$$s_1 = -\alpha + \sqrt{\alpha^2 - \omega_0^2} \tag{3.122}$$

$$s_2 = -\alpha - \sqrt{\alpha^2 - \omega_0^2}. \tag{3.123}$$

In Equations 3.122 and 3.123, α is the damping coefficient, and ω_0 is the resonant frequency. At resonant frequency, the reactive components cancel each other. The damping coefficient and resonant frequency are given by the following equations:

$$\alpha = \frac{1}{2RC} \tag{3.124}$$

and

$$\omega_0 = \frac{1}{\sqrt{LC}}. \tag{3.125}$$

When

$$\omega_0^2 < \alpha^2, \quad s_1 \text{ and } s_2 \quad \text{are real and distinct, voltage is overdamped}$$

$$\omega_0^2 > \alpha^2, \quad s_1 \text{ and } s_2 \quad \text{are compelx, voltage is underdamped} \tag{3.126}$$

$$\omega_0^2 = \alpha^2, \quad s_1 \text{ and } s_2 \quad \text{real and equal, voltage is critically damped}$$

The time domain representation of a parallel resonant network voltage response to illustrate underdamped and overdamped cases are illustrated in Figure 3.35a and b, respectively. L and C values are taken to be 0.1 H and 0.001 F, respectively, for an underdamped case, whereas for an overdamped case, L and C values are taken to be 50 mH and 0.2 μF, respectively. R values are varied to see its effect on voltage response for damping.

The quality factor and the bandwidth of the parallel resonant network are

$$Q = \frac{R}{\omega_0 L} = \omega_0 RC \tag{3.127}$$

$$BW = \frac{\omega_0}{Q} = \frac{1}{RC}. \tag{3.128}$$

(a)

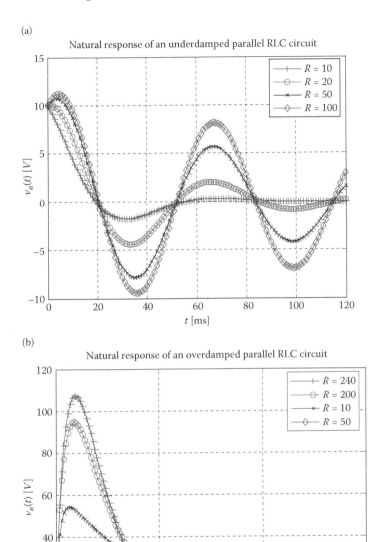

FIGURE 3.35 (a) Parallel resonant network response for underdamped case and (b) Parallel resonant network response for overdamped case.

In terms of quality factor, the roots given by Equations 3.122 and 3.123 can be expressed as

$$s_1 = \omega_0 \left[-\frac{1}{2Q} + \sqrt{\left(\frac{1}{2Q}\right)^2 - 1} \right]$$

(3.129)

$$s_2 = \omega_0 \left[-\frac{1}{2Q} - \sqrt{\left(\frac{1}{2Q}\right)^2 - 1} \right].$$

(3.130)

Assume now there is a source current connected to parallel resonant network in Figure 3.34 as illustrated in Figure 3.36.

The equivalent impedance of the parallel resonant network is found from Figure 3.36 as

$$Z_{eq}(s) = \frac{V_0(s)}{I_s(s)} = \frac{s/C}{s^2 + s(1/RC) + (1/LC)} = \frac{s/C}{(s - s_1)(s + s_2)},$$

(3.131a)

which can be written as

$$Z_{eq}(j\omega) = \frac{1}{R} + j\frac{\omega L}{(1 - \omega^2 LC)}.$$

(3.131b)

In Equation 3.131, s_1 and s_2 are now the poles of the impedance. When

$$\left(\frac{1}{2RC}\right)^2 \geq \left(\frac{1}{LC}\right) \text{ or } R \leq \left(\frac{\omega_0 L}{2}\right) = \left(\frac{1}{2\omega_0 C}\right)$$

(3.132)

poles of the impedance lie on the negative real axis. Hence, the value of R is small when compared to the values of the reactances, and as a result, the resonant network has a broadband response. When

$$\left(\frac{1}{2RC}\right)^2 < \left(\frac{1}{LC}\right) \text{ or } R > \left(\frac{\omega_0 L}{2}\right),$$

(3.133)

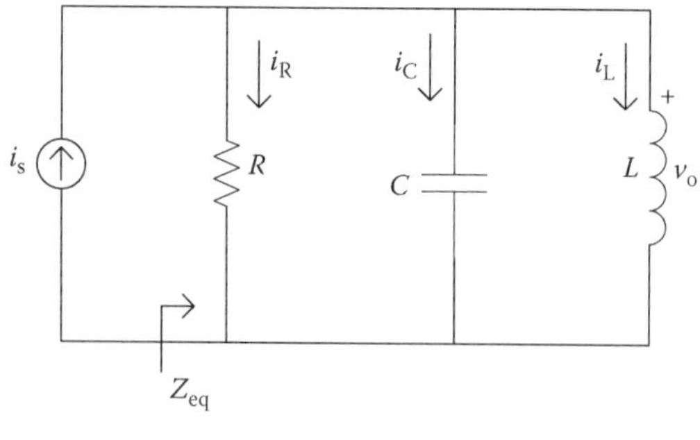

FIGURE 3.36 Parallel resonant circuit with source current.

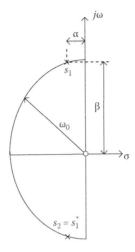

FIGURE 3.37 Pole-zero diagram for complex conjugate roots.

poles become complex, and they take the following form:

$$s_1 = -\alpha + j\sqrt{\omega_0^2 - \alpha^2} = -\alpha + j\beta \qquad (3.134)$$

$$s_2 = -\alpha - j\sqrt{\omega_0^2 - \alpha^2} = -\alpha - j\beta. \qquad (3.135)$$

This can be illustrated on the pole-zero diagram as shown in Figure 3.37.

The transfer function for the parallel resonant network is found using Figure 3.34 as

$$|H(\omega)| = \left|\frac{I_R}{I_S}\right| = \frac{\omega(L/R)}{\sqrt{\left(1 - \omega^2 LC\right)^2 + \left(\omega L/R\right)^2}}. \qquad (3.136)$$

At the resonant frequency, the transfer function will be real and be equal to its maximum value as

$$|H(\omega = \omega_0)| = \frac{\omega_0(L/R)}{\sqrt{\left(1 - \omega_0^2 LC\right)^2 + \left(\omega_0 L/R\right)^2}} = 1 = |H(\omega)|_{max}. \qquad (3.137)$$

The network response is obtained using transfer function given by Equation 3.137 for different values of R as shown in Figure 3.38. The values of the inductance and capacitance are taken to be $1.25\,\mu H$ and $400\,nF$, respectively.

This gives the resonant frequency as $0.22508\,MHz$. The condition for a broadband network is accomplished when $R = 5$ as shown in Figure 3.38. As R increases, the quality factor of the network increases, which agrees with the Equation 3.127.

Quality factor, Q, is an important parameter in resonant network response because it can be used as a measure for the loss and bandwidth of the circuit.

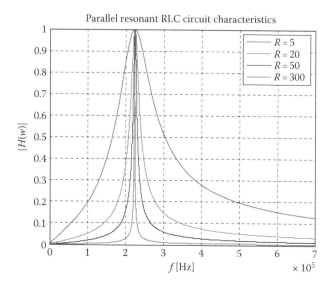

FIGURE 3.38 Parallel resonant circuit transfer function characteristics.

The quality factor of the circuit defines the ratio of the peak energy stored to the energy dissipated per cycle is given by

$$Q = \frac{2\pi \text{ (Peak energy stored)}}{\text{(Energy dissipated per cycle stored)}} = \frac{2\pi\left(\frac{1}{2}CV^2\right)}{(2\pi/\omega_0)(V^2/2R)} = \omega_0 CR. \qquad (3.138)$$

Example 3.9

A parallel resonant circuit with a source resistance of 50 Ω and a load resistance of 25 Ω. The loaded Q must be equal to 12 at the resonant frequency of 60 MHz.

1. Design the resonant circuit.
2. Calculate the 3dB bandwidth of the resonant circuit.
3. Use MATLAB to obtain the frequency response of this circuit versus frequency, i.e., plot $20\log(V_o/V_{in})$ versus frequency.

Solution:

1. The effective parallel resistance across parallel resonance circuit is

$$R_p = \frac{(50)25}{50+25} = 16.67[\Omega].$$

Then,

$$X_p = \frac{R_p}{Q} = \frac{16.67}{12} = 1.4.$$

Since,

$$X_p = \omega L = \frac{1}{\omega C}$$

then, the resonance element values are

$$L = \frac{X_p}{\omega} = \frac{1.4}{2\pi\left(60\times 10^6\right)} = 3.71\,[\text{nH}]$$

and

$$C = \frac{1}{\omega X_p} = \frac{1}{2\pi\left(60\times 10^6\right)(1.4)} = 1894.7\,[\text{pF}].$$

2. The BW is found from

$$Q = \frac{f_c}{BW_{3dB}} \rightarrow BW_{3dB} = \frac{f_c}{Q} = \frac{60\times 10^6}{12} = 5\times 10^6\,[\text{Hz}].$$

3. The MATLAB script to obtain the attenuation profile is given below.

```
clear
f = linspace(1,100*10^6);
RL = 25;
RS = 50;
RP = 50*25/(50+25);
Q = 12;
XP = RP/Q;

fc = 60*10^6;
wc = 2*pi()*fc;
L = XP/(wc);
C = 1/(wc*XP);

w = 2*pi.*f;
XL=1j.*w.*L;
XC=-1j./(w.*C);
Xeq=(XL.*XC)./(XL+XC);
Zeq=(RL.*Xeq)./(RL+Xeq);
S21=20.*log10(abs(Zeq./(Zeq+RS)));

plot(f,S21);

grid on
title('Attenuation Profile')
xlabel('Frequency (Hz)')
ylabel('Attenuation (dB)')
```

The plot of the attenuation profile is given in Figure 3.39.

3.7.1.2   Series Resonance

Consider the series resonant circuit given in Figure 3.40. The response of the circuit can be obtained by application of KVL, which gives the first-order differential equation for current i as

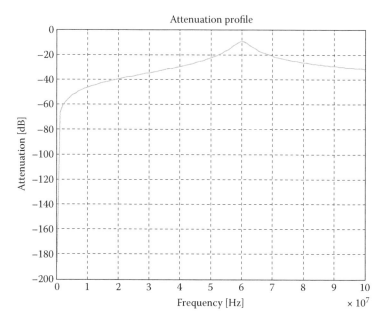

FIGURE 3.39 Attenuation profile versus frequency.

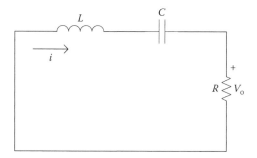

FIGURE 3.40 Series resonant network.

$$Ri + L\frac{di}{dt} + \frac{1}{C}\int_0^t i\,d\tau + V_o = 0, \tag{3.139}$$

where V_o is the initial charged voltage on capacitor. Equation 3.139 can be written as

$$\frac{d^2i}{dt^2} + \frac{R}{L}\frac{di}{dt} + \frac{i}{LC} = 0. \tag{3.140}$$

Following the same solution technique for parallel resonant network leads to the following equation in frequency domain:

$$s^2 + \frac{R}{L}s + \frac{1}{LC} = 0. \tag{3.141}$$

The roots of the equation are

$$s_1 = -\frac{R}{2L} + \sqrt{\left(\frac{R}{2L}\right)^2 - \left(\frac{1}{LC}\right)} \tag{3.142}$$

$$s_1 = -\frac{R}{2L} - \sqrt{\left(\frac{R}{2L}\right)^2 - \left(\frac{1}{LC}\right)}. \tag{3.143}$$

Equations 3.142 and 3.143 can be written as

$$s_1 = -\alpha + \sqrt{\alpha^2 - \omega_o^2} \tag{3.144}$$

$$s_2 = -\alpha - \sqrt{\alpha^2 - \omega_o^2}, \tag{3.145}$$

where α and ω are defined for series resonant network as

$$\alpha = \frac{R}{2L} \tag{3.146}$$

and

$$\omega_0 = \frac{1}{\sqrt{LC}}. \tag{3.147}$$

The quality factor and the bandwidth of the series resonant network are

$$Q = \frac{\omega_0 L}{R} = \frac{1}{\omega_0 RC} \tag{3.148}$$

$$BW = \frac{\omega_0}{Q} = \frac{R}{L}. \tag{3.149}$$

In terms of quality factor, the roots given by Equations 3.142 and 3.143 can be obtained as

$$s_1 = \omega_0 \left[-\frac{1}{2Q} + \sqrt{\left(\frac{1}{2Q}\right)^2 - 1} \right] \tag{3.150}$$

$$s_2 = \omega_0 \left[-\frac{1}{2Q} - \sqrt{\left(\frac{1}{2Q}\right)^2 - 1} \right]. \tag{3.151}$$

Equations 3.150 and 3.151 are identical to the ones obtained for parallel resonant circuit. The damping characteristics of the series resonant network follow the conditions listed by Equation 3.126. The time domain representation of a series resonant network current response illustrating underdamped and overdamped cases are illustrated in Figures 3.41 and 3.42, respectively. L and C values are taken to be $100\,\text{mH}$ and $10\,\mu\text{F}$ for an underdamped case, whereas for overdamped case, L and C values are taken to be $200\,\text{mH}$ and $10\,\mu\text{F}$. R values are varied to see its effect on current response for damping.

The transfer function of this network can be found by connecting a source voltage as shown in Figure 3.43 and is obtained as

$$|H(j\omega)| = \left|\frac{V_o(s)}{V_s(s)}\right| = \frac{\omega\frac{R}{L}}{\sqrt{\left(\frac{1}{LC} - \omega^2\right)^2 + \left(\omega\frac{R}{L}\right)^2}}. \tag{3.152}$$

The phase of the transfer function is found from

$$\theta(j\omega) = 90^\circ - \tan^{-1}\left(\frac{\omega\frac{R}{L}}{\frac{1}{LC} - \omega^2}\right) \tag{3.153}$$

At resonant frequency, the transfer function is maximum and will be equal to

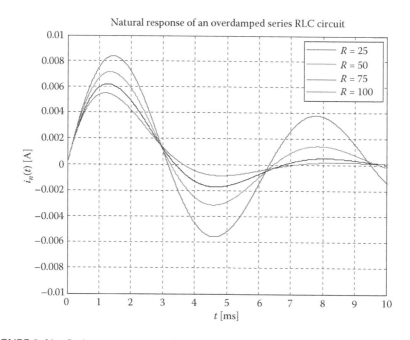

FIGURE 3.41   Series resonant network response for underdamped case.

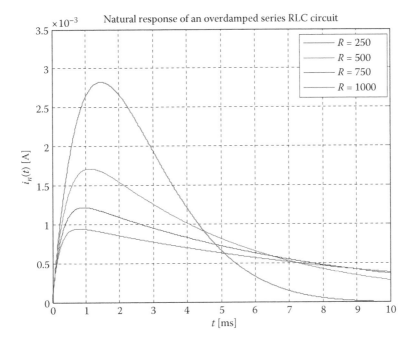

FIGURE 3.42 Series resonant network response for overdamped case.

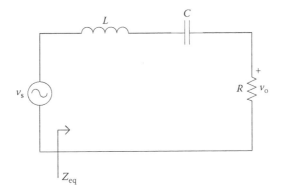

FIGURE 3.43 Series resonant network with source voltage.

$$\left|H(j\omega)\right| = \frac{\sqrt{\frac{1}{LC}}\frac{R}{L}}{\sqrt{\left(\frac{1}{LC} - \frac{1}{LC}\right)^2 + \left(\sqrt{\frac{1}{LC}}\frac{R}{L}\right)^2}} = 1 = \left|H(j\omega)\right|_{\max}. \qquad (3.154)$$

The resonant characteristics of the network can be obtained by plotting the transfer function given by Equation 3.152 versus different values of R as shown in Figure 3.44. The values of the inductance and capacitance are taken to be 150 μH and 10 nF.

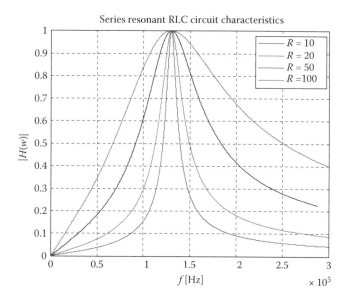

FIGURE 3.44 Series resonant circuit transfer function characteristics.

3.7.2 PARALLEL *LC* NETWORKS

3.7.2.1 Parallel *LC* Networks with Ideal Components

When ideal components are used, the typical circuit would be represented as the one shown in Figure 3.45. At resonance, the magnitudes of the reactances of the *L* and *C* elements are equal. The reactances of two components have opposite signs; thus, the net reactance is zero for a series circuit or infinity for a parallel circuit. Hence, we can obtain the resonance frequency from

$$\omega_0 L = \frac{1}{\omega_0 C} \rightarrow \omega_0 = \frac{1}{\sqrt{LC}} \rightarrow f_0 = \frac{1}{2\pi\sqrt{LC}}. \tag{3.155}$$

The transfer function for *LC* resonant network in Figure 3.45 is found from

$$|H(\omega)|_{dB} = \left|\frac{V_{out}}{V_{in}}\right|_{dB} = 20\log\left(\frac{X_{total}}{R_s + X_{total}}\right), \tag{3.156}$$

where

$$X_{total} = \frac{X_C X_L}{X_C + X_L}. \tag{3.157}$$

Substituting Equation 3.157 into 3.156 gives

$$|H(\omega)|_{dB} = \left|\frac{V_{out}}{V_{in}}\right|_{dB} = 20\log\left|\frac{j\omega L / R_s}{\left(1 - \omega^2 LC\right) + j\omega L / R_s}\right|. \tag{3.158}$$

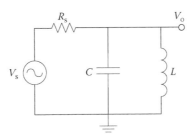

FIGURE 3.45 *LC* resonant network with ideal components and source.

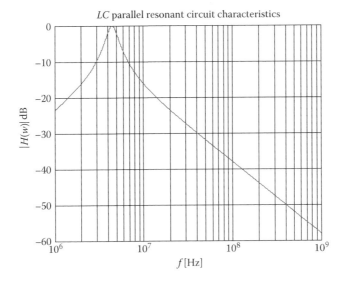

FIGURE 3.46 The frequency characteristics of *LC* network with source.

The frequency characteristics of the circuit are plotted in Figure 3.46 when $L = 0.5\,\mu\text{H}$, $C = 2500\,\text{pF}$, and $R = 50\,\Omega$.

The quality factor of the circuit is found from its equivalent impedance as

$$Z_{eq} = R_s + \frac{X_C X_L}{X_C + X_L} = \frac{R_s}{\left(1 - \omega^2 LC\right)} + j\frac{\omega L}{\left(1 - \omega^2 LC\right)}. \tag{3.159}$$

Then, the quality factor of the circuit is

$$Q = \frac{|X|}{R} = \frac{\omega L}{R_s}. \tag{3.160}$$

3.7.2.2 Parallel *LC* Networks with Nonideal Components

Now, assume that we have some additional loss for inductor for the *LC* network as shown in Figure 3.47.

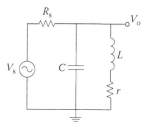

FIGURE 3.47   Addition of loss to parallel LC network.

The equivalent impedance of the network in Figure 3.47 takes the following form with the addition of loss component, r:

$$Z_{eq} = R_s + \frac{r + j\omega L}{\left(j\omega rC - \omega^2 LC + 1\right)} = R_s + \frac{\left(r + j\omega L\right)\left(1 - \omega^2 LC - j\omega rC\right)}{\left(1 - \omega^2 LC\right)^2 + \left(\omega rC\right)^2}, \quad (3.161)$$

which can be simplified to

$$Z_{eq} = \frac{R_s\left[\left(1 - \omega^2 LC\right)^2 + \left(\omega rC\right)^2\right] + r}{\left(1 - \omega^2 LC\right)^2 + \left(\omega rC\right)^2} + j\frac{\omega\left[\left(L - Cr^2\right) - \omega^2 L^2 C\right]}{\left(1 - \omega^2 LC\right)^2 + \left(\omega rC\right)^2}. \quad (3.162)$$

The resonant frequency of the network is now equal to

$$\omega_0 = \sqrt{\frac{L - Cr^2}{L^2 C}} \rightarrow f_0 = \frac{1}{2\pi}\sqrt{\frac{L - Cr^2}{L^2 C}}. \quad (3.163)$$

The loaded quality factor of the circuit is obtained as

$$Q_L = \frac{\omega\left[\left(L - Cr^2\right) - \omega^2 L^2 C\right]}{R_s\left[\left(1 - \omega^2 LC\right)^2 + \left(\omega rC\right)^2\right]}. \quad (3.164)$$

When $r = 0$ for the lossless case, Equations 3.163 and 3.164 reduce to the ones given in Equations 3.155 and 3.160. The transfer function with the loss resistance changes to

$$|H(\omega)|_{dB} = \left|\frac{V_{out}}{V_{in}}\right|_{dB} = 20\log\left|\frac{r + j\omega\left[\left(L - Cr^2\right) - \omega^2 L^2 C\right]}{R_s\left[\left(1 - \omega^2 LC\right)^2 + \left(\omega rC\right)^2\right] + j\omega\left[\left(L - Cr^2\right) - \omega^2 L^2 C\right]}\right|.$$

$$(3.165)$$

The transfer function showing the attenuation profile with different loss resistance values when $L = 0.5\,\mu H$, $C = 2500\,pF$, and $R = 50\,\Omega$ are illustrated in Figure 3.48. The network bandwidth broadens as the value of r increases as expected.

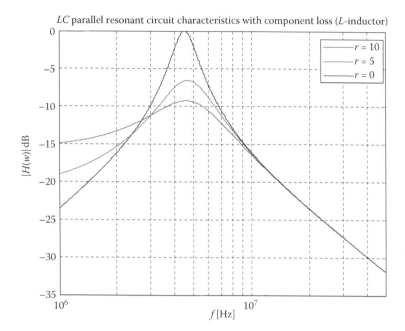

FIGURE 3.48 Attenuation profile of *LC* network with loss resistor.

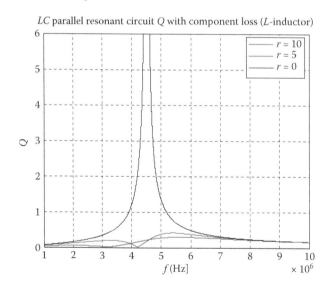

FIGURE 3.49 Quality factor of *LC* network with loss resistor.

The quality factor of the network with the addition of loss resistance significantly differs from the original *LC* parallel resonant network. Original network has ideally infinite value of quality factor. In agreement with this, a very large value of the quality factor for the original network is obtained at resonance frequency when $r = 0$ is also seen in Figure 3.49.

3.7.2.3 Loading Effects on Parallel *LC* Networks

Resonant circuit becomes loaded when it is connected to a load or when fed by a source Q of the circuit under these conditions called loaded Q or simply Q_L. The loaded of the circuit then depends on source resistance, load resistance, and individual Q of the reactive components.

When the reactive components are lossy, they affect the Q factor of the overall circuit. For instance, consider the resonant circuit with source resistance in Figure 3.45. The quality factor of the circuit for various source resistance values when $L = 0.5\,\mu H$, $C = 2500\,pF$ has been illustrated in Figure 3.50. For the same frequency, the quality factor increases as the value of source resistance, R_s, increases. As a result, the selectivity of the network can be adjusted by setting the value of source resistance. The attenuation profile showing the response when source resistance is changed is given in Figure 3.51.

The circuit shown in Figure 3.52 has both source and load resistances. The impedance of the loaded resonant network can be obtained as

$$Z_{eq} = R_s + \frac{j\omega R_L L}{\omega L + j\omega R_L \left(\omega^2 LC - 1\right)} = R_s + \frac{\omega R_L L\left[\omega L - j\omega R_L \left(\omega^2 LC - 1\right)\right]}{\left(\omega L\right)^2 + \left[\omega R_L \left(\omega^2 LC - 1\right)\right]^2}, \quad (3.166)$$

which can be simplified to

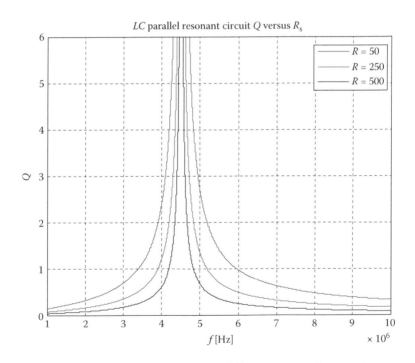

FIGURE 3.50 Quality factor of *LC* network for different source resistance values.

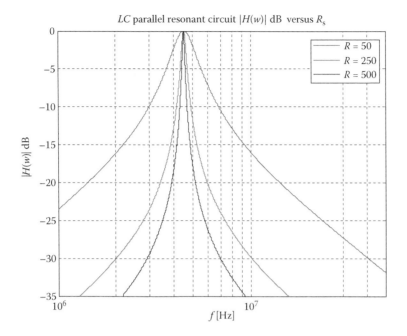

FIGURE 3.51   Attenuation profile of LC network for different source resistance values.

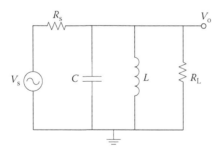

FIGURE 3.52   Loaded LC resonant circuit.

$$Z_{eq} = \frac{R_s\left(\left(\omega L\right)^2 + \left[\omega R_L\left(\omega^2 LC - 1\right)\right]^2\right) + \left(\omega L\right)^2 R_L}{\left(\omega L\right)^2 + \left[\omega R_L\left(\omega^2 LC - 1\right)\right]^2} + j\frac{\omega R_L^2 L\left(1 - \omega^2 LC\right)}{\left(\omega L\right)^2 + \left[\omega R_L\left(\omega^2 LC - 1\right)\right]^2}.$$

(3.167)

The loaded quality factor of the circuit is obtained as

$$Q_L = \frac{\omega R_L^2 L\left(1 - \omega^2 LC\right)}{R_s\left(\left(\omega L\right)^2 + \left[\omega R_L\left(\omega^2 LC - 1\right)\right]^2\right) + \left(\omega L\right)^2 R_L}.$$

(3.168)

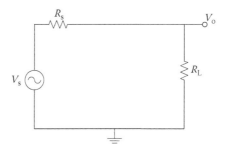

FIGURE 3.53   Equivalent loaded LC resonant circuit at resonance.

At resonant frequency, this circuit is simplified to the one in Figure 3.53.
   The equivalent impedance from Equation 3.167 is equal to

$$Z_{eq} = R_s + R_L.$$ (3.169)

It is important to note that at resonant frequency,

$$\omega_0 L = \frac{1}{\omega_0 C}.$$ (3.170)

3.8 LC RESONATORS AS IMPEDANCE TRANSFORMERS

3.8.1   INDUCTIVE LOAD

Consider the LC parallel network with a loss resistor. The new circuit can be illustrated in Figure 3.54.
   The equivalent impedance at the input for the circuit in Figure 3.54 can be written as

$$Z_{eq} = \frac{R}{\left(1-\omega^2 LC\right)^2 + \left(\omega RC\right)^2} + j\frac{\omega\left[\left(L - CR^2\right) - \omega^2 L^2 C\right]}{\left(1-\omega^2 LC\right)^2 + \left(\omega RC\right)^2}.$$ (3.171)

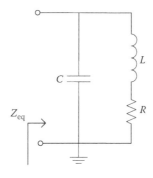

FIGURE 3.54 LC impedance transformer for inductive load.

The resonant frequency of the network is now equal to

$$\omega_0 = \sqrt{\frac{L-CR^2}{L^2C}} \rightarrow f_0 = \frac{1}{2\pi}\sqrt{\frac{L-CR^2}{L^2C}}. \tag{3.172}$$

At resonance frequency, the equivalent impedance will be purely resistive, $Z_{eq} = R_{eq}$, and equal to

$$R_{eq} = \frac{R}{\left(1-\omega^2LC\right)^2+\left(\omega RC\right)^2} = \frac{L}{RC}. \tag{3.173}$$

Hence, when the network is at resonance, the capacitive load impedance is converted into a resistive input impedance. The quality factor, Q_{load}, of the load at resonance is

$$Q_{load} = \frac{\omega_0 L}{R}. \tag{3.174}$$

The following relation between the load quality factor and the equivalent impedance at resonance can be written as

$$R_{eq} = \frac{L}{RC} = \left(Q_{load}^2+1\right)R. \tag{3.175}$$

3.8.2 CAPACITIVE LOAD

The same principle for inductive load can be applied to convert the capacitive load to a resistive load at the resonant frequency using the LC resonant circuit shown in Figure 3.55.

The equivalent impedance at the input for the circuit in Figure 3.55 can be written as

$$Z_{eq} = \frac{\omega^4 RL^2C^2}{\left(1-\omega^2LC\right)^2+\left(\omega RC\right)^2} + j\frac{\omega L\left[\omega^2\left(R^2C^2-LC\right)+1\right]}{\left(1-\omega^2LC\right)^2+\left(\omega RC\right)^2}. \tag{3.176}$$

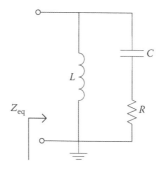

FIGURE 3.55 LC impedance transformer for capacitive load.

The resonant frequency of the network is now equal to

$$\omega_0 = \sqrt{\frac{1}{LC - R^2 C^2}} \rightarrow f_0 = \frac{1}{2\pi} \sqrt{\frac{1}{LC - R^2 C^2}}. \qquad (3.177)$$

At resonance frequency, $Z_{eq} = R_{eq}$, and it can be expressed as

$$R_{eq} = \frac{L}{RC}. \qquad (3.178)$$

Hence, when the network is at resonance, the capacitive load impedance is converted into a resistive input impedance. Since the quality factor, Q_{load}, of the load at resonance is

$$Q_{load} = \frac{1}{\omega_0 RC}, \qquad (3.179)$$

then the following relation can be established:

$$R_{eq} = \frac{L}{RC} = \omega_0 L Q_{load}. \qquad (3.180)$$

Example 3.10

An amplifier output needs to be terminated with a load line resistance of 2000 Ω at 1.6 MHz. It is given in the data sheet that the transistor has 20 pF at 1.6 MHz. There is an inductive load connected to the output of the load line circuit of the amplifier with $R_L = 5\ \Omega$. The configuration of this circuit is given in Figure 3.56.

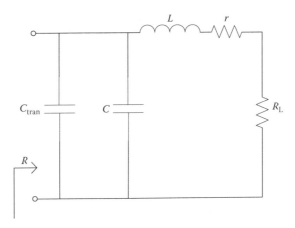

FIGURE 3.56 Amplifier output load line circuit.

1. Calculate the values of L and C by assuming that the load inductor has a negligible loss, i.e., $r = 0$.
2. The inductor is changed to a magnetic core inductor, which has the quality factor of 50. Calculate loss resistance, r, for the reactive component values obtained in (1). What is the value of new load line resistance?
3. If the quality factor of the inductor is 50 and load line resistor is required to be 2000 Ω as set in the problem, what are the values of L and C with $R_L = 5\ \Omega$?

Solution: It is given that $R_{eq} = 2000\ \Omega$, $R_L = R = 5\ \Omega$, $C_{tran} = 20\,pF$, $f = 1.6\,MHz$, $\omega_0 = 10^7\,rad/s$, and $C_T = C_{tran} + C$.

1. When $r = 0\ \Omega$,

$$R_{eq} = \left(Q_{load}^2 + 1\right)R \rightarrow \frac{R_{eq}}{R} - 1 = Q_{load}^2 \rightarrow Q_{load} = 19.98. \tag{3.181}$$

So,

$$Q_{load} = \frac{\omega_0 L}{R} \rightarrow L = \frac{Q_{load}R}{\omega_0} \rightarrow L = 10\,[\mu H]. \tag{3.182}$$

Hence,

$$R_{eq} = \frac{L}{RC_T} \rightarrow C_T = \frac{L}{R_{eq}R} \rightarrow C_T = 1\,[nF], \tag{3.183}$$

where

$$C_T = C_{tran} + C \rightarrow C = C_T - C_{tran} \rightarrow C = 980\ [pF]. \tag{3.184}$$

2. The Q of the inductor is given to be equal to 50. Then,

$$Q_{inductor} = \frac{\omega_0 L}{r} \rightarrow r = \frac{\omega_0 L}{Q_{inductor}} \rightarrow r = 2\,[\Omega]. \tag{3.185}$$

So, the new load resistance, $R_{eq'}$ is

$$R_{eq} = \frac{L}{(R+r)C_T} = \frac{10 \times 10^{-6}}{(7)(1 \times 10^{-9})} = 1428.6\ [\Omega]. \tag{3.186}$$

3. The Q of the inductor is given to be equal to 50. Then,

$$Q_{inductor} = \frac{\omega_0 L}{r} \rightarrow r = 2 \times 10^5 L = aL. \tag{3.187}$$

Since,

$$R_{eq} = R\left(Q_{load}^2 + 1\right) \rightarrow R_{eq} = (R+r)\left(Q_{load}^2 + 1\right) = \frac{\omega_0^2 L^2 + (R + aL)^2}{R + aL}, \tag{3.188}$$

which leads to solution for L as

$$L^2 - L\frac{a(R_{eq} - 2R)}{\omega_0^2 + a^2} - \frac{R(R_{eq} - R)}{\omega_0^2 + a^2} = 0. \tag{3.189}$$

As a result, the value of r is found as

$$r = 2 \times 10^5 L = 2.44 \ [\Omega], \tag{3.190}$$

where

$$L = 12.2 \ [\mu H]$$

Using Equation 3.183,

$$R_{eq} = \frac{L}{RC_T} \rightarrow C_T = \frac{L}{R_{eq}R} \rightarrow C_T = 820 \ [pF], \tag{3.191}$$

where

$$C_T = C_{tran} + C \rightarrow C = 820 - 20 \rightarrow C = 800 \ [pF]. \tag{3.192}$$

3.9 TAPPED RESONATORS AS IMPEDANCE TRANSFORMERS

3.9.1 TAPPED-C IMPEDANCE TRANSFORMER

To understand the operation of tapped-C impedance transformer, consider the capacitive voltage divider circuit shown in Figure 3.57. The output voltage can be found from

$$v_o = v_i \frac{1/(j\omega C_2)}{1/(j\omega C_2) + 1/(j\omega C_1)} = v_i \frac{C_1}{C_1 + C_2}, \tag{3.193}$$

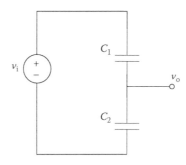

FIGURE 3.57 Capacitive voltage divider.

which can be expressed as

$$v_o = v_i n,$$

where

$$n = \frac{C_1}{C_1 + C_2}. \qquad (3.194)$$

Now, assume there is a load resistor connected to the output of the capacitor and resonator inductor connected to the input of the divider circuit as shown in Figure 3.58.

We then convert the output shunt-connected circuit to a series connection by using parallel-to-series conversion introduced earlier as shown in Figure 3.59.

The relation of the components in Figures 3.58 and 3.59 are

$$C_s = C_2 \left(\frac{Q_p^2 + 1}{Q_p^2} \right) \qquad (3.195)$$

$$R_s = \frac{R}{Q_p^2 + 1} \qquad (3.196)$$

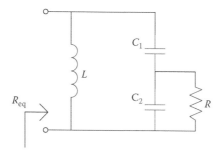

FIGURE 3.58 Capacitive voltage divider with load resistor.

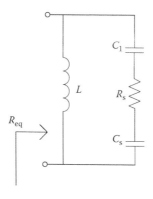

FIGURE 3.59 Capacitive voltage divider with parallel-to-series transformation.

$$R_s = \frac{R_{eq}}{Q_r^2 + 1}, \tag{3.197}$$

where

$$Q_p = \frac{R}{X_{C_2}} = \omega_0 R C_2 \tag{3.198}$$

$$Q_r = \frac{R_{eq}}{\omega_0 L} = \frac{1}{\omega_0 R_s C}. \tag{3.199}$$

The equivalent capacitance can then be written as

$$C = \frac{C_1 C_s}{C_1 + C_s}. \tag{3.200}$$

Equating 3.195 and 3.198 gives

$$Q_p = \sqrt{\left[(Q_r^2 + 1) \frac{R}{R_{eq}} - 1 \right]}. \tag{3.201}$$

Overall, using the transformations given, the tapped-C circuit in Figure 3.60a can be simplified and transformed to the one in Figure 3.60b with the following relations as

$$R_s' = R_s \left(1 + \frac{C_1}{C_2} \right)^2 \tag{3.202}$$

and

$$C_T = \frac{C_1 C_2}{C_1 + C_2}. \tag{3.203}$$

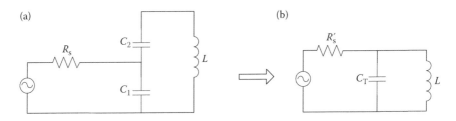

FIGURE 3.60 Tapped equivalent circuit.

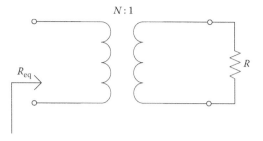

FIGURE 3.61 Equivalent tapped-C circuit representation using transformer.

At resonance, the circuit can be simplified to the impedance transformer circuit shown in Figure 3.61 with

$$N^2 = \frac{R}{R_{eq}}$$ (3.204)

Substitution of Equations 3.202 and 3.203 into 3.201 gives

$$Q_p = \sqrt{\left[\frac{\left(Q_r^2+1\right)}{N^2}-1\right]}.$$ (3.205)

Example 3.11

Design a parallel resonant circuit with the tapped-C approach where 3 dB bandwidth is 3 MHz and center frequency is 27.12 MHz. Resonant circuit will operate between a source resistance of 50 Ω and a load resistance of 100 Ω. Assume that Q of the inductor is 150 at 27.12 MHz.

1. Obtain element values of the circuit shown in Figure 3.62a.
2. Obtain element values of the equivalent circuit shown in Figure 3.62b.
3. Use MATLAB to obtain the frequency response of the circuits shown in Figure 3.62a and b.

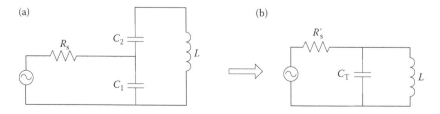

FIGURE 3.62 Parallel resonant circuit with tapped-C approach.

Solution: The equivalent circuit shown in Figure 3.54b has

$$R'_s = 100 \ \Omega$$

since

$$R'_s = R_s \left(1 + \frac{C_1}{C_2}\right)^2 \to \left(1 + \frac{C_1}{C_2}\right)^2 = 2 \to C_1 = 0.414 C_2.$$

Since the inductor is lossy,

$$X_p = \frac{R_p}{Q_p} \to R_p = Q_p X_p = 150 X_p.$$

The loaded Q of the resonant circuit is found from

$$Q = \frac{f_c}{f_2 - f_1} = \frac{27.12}{3} = 9.04.$$

Since

$$Q = \frac{R_{total}}{X_p} \to 9.04 = \frac{R_{total}}{X_p} = \frac{50 R_p}{(50 + R_p) X_p} \to 9.04 = \frac{(150 X_p) 50}{(50 + 150 X_p) X_p},$$

then,

$$X_p = \frac{7048}{1356} = 5.2 \ [\Omega].$$

So,

$$R_p = 150 X_p = 780 \ [\Omega].$$

The values of the L and C are found from

$$L = \frac{X_p}{\omega} = \frac{5.2}{2\pi (27.12 \times 10^6)} = 30.5 \ [nH]$$

$$C_T = \frac{1}{\omega X_p} = \frac{1}{2\pi (27.12 \times 10^6)(5.2)} = 1128 \ [pF].$$

The capacitor values for the circuit in Figure 3.54a are found from

$$C_T = \frac{C_1 C_2}{C_1 + C_2} \to 1128 = \frac{0.414 C_2}{1.414} \to C_2 = 3852.6 \ [pF] \text{ and } C_1 = 1595 \ [pF].$$

3. The attenuation profile is shown in Figure 3.63.

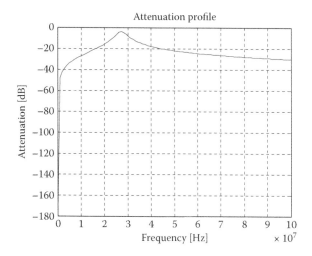

FIGURE 3.63 Attenuation profile for parallel resonant circuit with tapped-C.

3.9.2 Tapped-L Impedance Transformer

Typical tapped-L impedance transformer circuit is shown in Figure 3.64a. We can follow the same procedure outlined in Section 3.8.1 and convert the circuit to its equivalent circuit shown in Figure 3.64b by using parallel-to-series transformation relations as previously done.

Overall, the tapped-L circuit in Figure 3.64a can be simplified to the one shown in Figure 3.65 using the transformations with the following relation as

$$R'_s = R_s \left(\frac{n}{n_1} \right)^2. \tag{3.206}$$

3.10 SIGNAL FLOW GRAPHS

A signal flow graph technique is used to facilitate analysis of transmission lines in the amplifier design by providing a simplification for the complicated circuits. They are used to determine the critical amplifier design parameters such as reflection coefficients, power, and voltage gains. When signal flow graph of the circuit is obtained, mathematical relations are developed using Mason's rule. The key elements for the signal flow graph are as follows:

- Each variable is treated as a node.
- Branches represent paths for signal flow.
- The network must be linear.

A node represents the sum of the branches coming into it. Branches are represented by scattering parameters. It is safe to assume that branches enter dependent variable

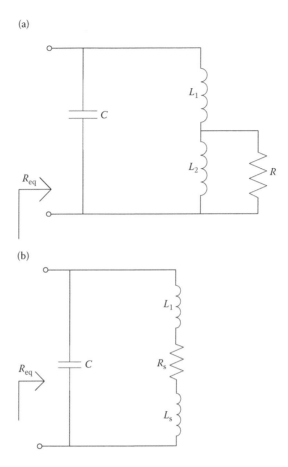

(a)

(b)

FIGURE 3.64   (a) Tapped-L impedance transformer and (b) tapped-L impedance transformer with parallel-to-series transformation.

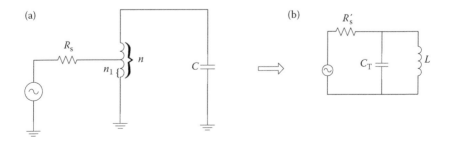

(a)

(b)

FIGURE 3.65   Tapped-L equivalent circuit.

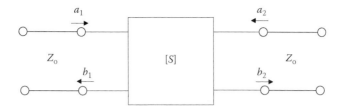

FIGURE 3.66   Two-port linear network illustration.

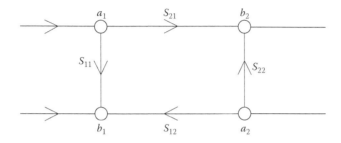

FIGURE 3.67   Signal flow graph implementation of two-port network.

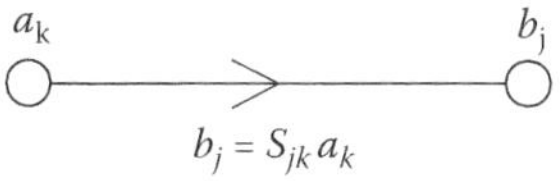

$$a_k \xrightarrow{\quad} b_j$$
$$b_j = S_{jk} a_k$$

FIGURE 3.68   Representation of branch using scattering parameter.

nodes and leave independent variable nodes. Consider the two-port linear network given in Figure 3.66. This network can be represented using the signal flow graph as shown in Figure 3.67. Scattering parameter on each branch is represented by the ratio of the reflected wave to the incident wave:

$$S_{jk} = \frac{b_j}{a_k}. \qquad (3.207)$$

This can be illustrated with the signal flow graph shown in Figure 3.68 where the wave that is emanated at the node a_k is assumed to be the incident wave and the wave that goes into node b_j is assumed to be the reflected wave.

Example 3.12

If signal is given as

$$b = S_{11}a_1 + S_{12}a_2,$$

find its signal flow graph representation?

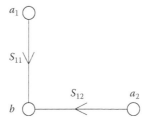

FIGURE 3.69 Representation of dependent node.

Solution: Signal b is a dependent node and can be represented as the two incoming branches as shown in Figure 3.69.

Example 3.13

Represent the signal source and source impedance given in Figure 3.70 by a signal flow graph.

Solution: Using the circuit in Figure 3.70, we can write the expression for the voltage at the input as

$$V_i = V_s + I_g Z_s,$$ (3.208)

which can be written in terms of incident and reflected wave as

$$V_i^+ + V_i^- = V_s + \left(\frac{V_i^+}{Z_0} - \frac{V_i^-}{Z_0} \right) Z_s.$$ (3.209)

Solving Equation 3.209 for V_i^- gives

$$b_g = b_s + \Gamma_s a_g,$$ (3.210)

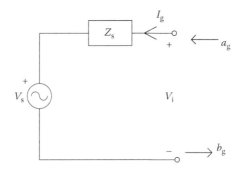

FIGURE 3.70 Source generator and impedance circuit.

where

$$b_g = \frac{V_i^-}{\sqrt{Z_0}}$$ (3.211)

$$a_g = \frac{V_i^+}{\sqrt{Z_0}}$$ (3.212)

$$b_s = \frac{V_s \sqrt{Z_0}}{Z_s + Z_0}.$$ (3.213)

Hence, we obtain

$$\Gamma_s = \frac{Z_s - Z_0}{Z_s + Z_0}.$$ (3.214)

The results can be represented with the signal flow graph shown in Figure 3.71.

Example 3.14

Represent the load impedance given in Figure 3.72 by a signal flow graph.

Solution: In Figure 3.72, the load voltage is represented as

$$V_L = Z_L I_L.$$ (3.215)

We can represent load voltage in terms of incident and reflected waves as

$$V_L^+ + V_L^- = Z_L + \left(\frac{V_L^+}{Z_0} - \frac{V_L^-}{Z_0} \right).$$ (3.216)

Equation 3.216 can be rewritten as

$$b_L = \Gamma_L a_L,$$ (3.217)

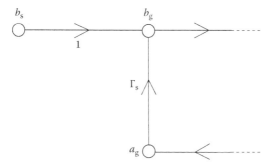

FIGURE 3.71 Signal flow graph representation of source generator and impedance circuit.

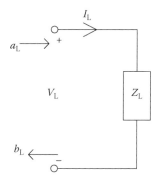

FIGURE 3.72 Load impedance circuit.

where

$$b_L = \frac{V_L^-}{\sqrt{Z_0}} \qquad (3.218)$$

$$a_L = \frac{V_L^+}{\sqrt{Z_0}} \qquad (3.219)$$

$$\Gamma_L = \frac{Z_L - Z_0}{Z_L + Z_0}. \qquad (3.220)$$

Using Equation 3.217, we can represent the load impedance in Figure 3.72 with the signal flow graph shown in Figure 3.73.

Example 3.15

Represent the two-port transmission circuit shown in Figure 3.74 with the signal flow graph.

Solution: We can now combine the solutions given in Figures 3.71 and 3.73 and obtain the signal flow graph representation for the transmission line circuit shown in Figure 3.74 as illustrated in Figure 3.75.

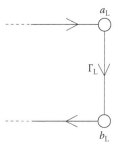

FIGURE 3.73 Signal flow graph representation of load impedance.

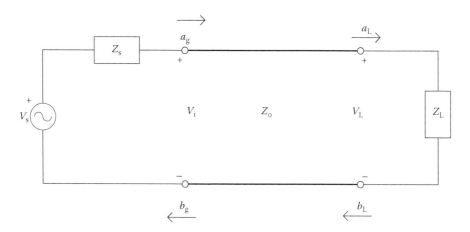

FIGURE 3.74   Transmission line circuit.

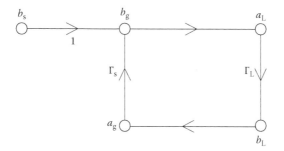

FIGURE 3.75   Signal flow graph representation of transmission line circuit.

REFERENCES

1. A. Eroglu, *RF Circuit Design Techniques for MF-UHF Applications*, CRC Press, Boca Raton, FL, 2013.
2. A. Eroglu, *Introduction to RF Power Amplifier Design and Simulation*, CRC press, Boca Raton, FL, 2016.

4 Small-Signal Amplifiers

4.1 AMPLIFIER BASIC TERMINOLOGY

In this section, some of the common terminologies used in radio frequency (RF) power amplifier (PA) design will be discussed. For this discussion, consider the simplified RF PA block diagram given in Figure 4.1. The RF PA shown in this figure has a simple three-port network comprising an RF input signal port, a DC input port, and an RF signal output port. Power that is not converted into RF output power, P_{out}, is dissipated as heat and designated as P_{diss} (Figure 4.1). The dissipated power, P_{diss}, is calculated as

$$P_{diss} = (P_{in} + P_{dc}) - P_{out} \tag{4.1}$$

4.1.1 GAIN

RF PA gain is defined as the ratio of the output power to the input power and is given as

$$G = \frac{P_{out}}{P_{in}} \tag{4.2}$$

It can be defined in terms of dB as

$$G[dB] = 10 \log\left(\frac{P_{out}}{P_{in}}\right)[dB] \tag{4.3}$$

RF PA gain is higher at lower frequencies. This can be illustrated based on the data measured for a switch-mode RF amplifier operating in the high-frequency (HF) range for various DC supply voltages (Figure 4.2).

It is possible to obtain higher gain level when multiple amplifiers are cascaded to obtain multistage amplifier configuration as shown in Figure 4.3.

The overall gain of the multistage amplifier system shown in Figure 4.3 can be given as

$$G_{tot}[dB] = G_{PA_1}[dB] + G_{PA_2}[dB] + G_{PA_3}[dB] \tag{4.4}$$

The unit of gain is given in terms of [dB], because it is the ratio of the output power to the input power. It is important to note that dB is not a unit to define the power. In amplifier terminology, dBm is used to define the power. dBm can be calculated as

$$dBm = 10 \log\left(\frac{P}{1\,mW}\right) \tag{4.5}$$

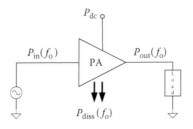

FIGURE 4.1   RF PA as three-port network.

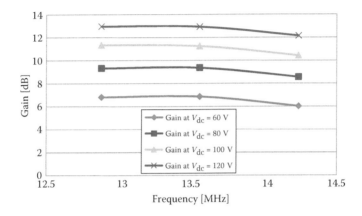

FIGURE 4.2   Measured gain variation versus frequency for a switched-mode RF amplifier.

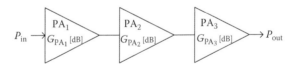

FIGURE 4.3   Multistage RF amplifiers.

Example 4.1

In the RF system shown in Figure 4.4, the RF signal source can provide power output from 0 to 30 dBm. RF signal is fed through a 1 dB T-pad attenuator and a 20 dB directional coupler, where the sample of the RF signal is further attenuated by a 3 dB π-pad attenuator before power meter reading in dB. The "through" port of the directional coupler has 0.1 dB of loss before it is sent to PA. RF PA output is then connected to a 6 dB π-pad attenuator. If power meter is reading 10 dBm, what is the power delivered to the load shown in Figure 4.4 in mW?

Solution: We need to find the source power first. The loss from power meter to RF signal source is

$$\text{Loss from Power Meter to Source} = 3\,\text{dB} + 20\,\text{dB} + 1\,\text{dB} = 23\,\text{dB}$$

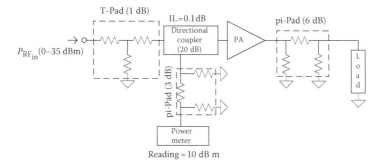

FIGURE 4.4 RF system with coupler and attenuation pads.

Then, the power at the source is

$$RF\ Source\ Signal = 23\ dBm + 10\ dBm = 33\ dBm$$

The total loss toward PA is due to the T-Pad attenuator (1 dB) and the directional coupler (0.1 dB) and is equal to 1.1 dB. Therefore, the transmitted RF signal at PA is

$$RF\ Signal\ at\ PA = 33\ dBm - 1.1\ dBm = 31.9\ dBm$$

Hence, the power delivered to the load is

$$Power\ Delivered\ to\ the\ Load = 31.9\ dBm - 6\ dBm = 25.9\ dBm$$

Power delivered in mW is

$$P[mW] = 10^{\frac{dBm}{10}} = 10^{\frac{25.9}{10}} = 389.04\ [mW]$$

4.1.2 Efficiency

In practical applications, RF PA is implemented as a subsystem, and it consumes most of the DC power from the supply. As a result, minimal DC power consumption for amplifier becomes important, and it can be accomplished by having high RF PA efficiency. RF PA efficiency is one of the critical and most important amplifier performance parameters. Amplifier efficiency can be used to define the drain efficiency for MOSFET or collector efficiency for bipolar junction transistor (BJT). Amplifier efficiency is defined as the ratio of the RF output power to the power supplied by DC source and can be expressed as

$$\eta[\%] = \frac{P_{out}}{P_{DC}} \times 100 \tag{4.6}$$

Efficiency in terms of gain can be expressed in the following form:

$$\eta[\%] = \frac{1}{1 + \left(\dfrac{P_{diss}}{P_{out}}\right) - \left(\dfrac{1}{G}\right)} \times 100 \tag{4.7}$$

The maximum efficiency is possible when there is no dissipation, i.e., $P_{diss} = 0$. The maximum efficiency calculated from Equation 4.7 is then equal to

$$\eta[\%] = \frac{1}{1 - \left(\dfrac{1}{G}\right)} \times 100 \tag{4.8}$$

When RF input power is included in the efficiency calculation, the efficiency is then called as power-added efficiency, η_{PAE}, and is calculated as

$$\eta_{PAE}[\%] = \frac{P_{out} - P_{in}}{P_{DC}} \times 100 \tag{4.9a}$$

or

$$\eta_{PAE}[\%] = \eta\left(1 - \frac{1}{G}\right) \times 100 \tag{4.9b}$$

Example 4.2

RF PA delivers 200 [W] to a given load. If the input supply power for this amplifier is 240 [W], and the power gain of the amplifier is 15 dB, find (1) drain efficiency and (2) power-added efficiency.

Solution:

1. The drain efficiency can be calculated from Equation 4.6 as

$$\eta[\%] = \frac{P_{out}}{P_{DC}} \times 100 = \frac{200}{240} \times 100 = 83.33\%$$

2. Power-added efficiency can be calculated from Equation 4.9 as

$$\eta_{PAE}[\%] = \eta\left(1 - \frac{1}{G}\right) \times 100 = 83\left(1 - \frac{1}{15}\right) = 77.47\%$$

4.1.3 POWER OUTPUT CAPABILITY

Power output capability of an amplifier is defined as the ratio of the output power of the amplifier to the maximum values of the voltage and current that device experiences during the operation of the amplifier. When more than one transistor is used, or the number of transistors increases due to the amplifier configuration used, such as push–pull configuration or any other combining techniques, this is reflected in the denominator of the following equation:

$$c_p = \frac{P_o}{N I_{max} V_{max}} \qquad (4.10)$$

4.1.4 LINEARITY

Linearity is a measure for RF amplifier output to follow the amplitude and phase of its input signal. In practice, the linearity of an amplifier is measured in a very different way. Linearity of an amplifier is measured by comparing the set power of an amplifier with the output power. The gain of the amplifier is then adjusted to compensate one of the closed-loop parameters. The typical closed-loop control system that is used to adjust the linearity of the amplifier through closed-loop parameters is shown in Figure 4.5. When the linearity of the amplifier is accomplished, the linear curve shown in Figure 4.6 is obtained. The experimental setup that is used to calibrate RF PAs to have linear characteristics is given in Figure 4.7.

In Figure 4.7, RF amplifier output is measured by thermocouple-based power meter via directional coupler. Directional coupler output is terminated with $50\,\Omega$ load. The set power is adjusted by the user, and the output forward power, P_{fwr}, and reverse power are measured with the power meter. If set power and output power are different, the control closed-loop parameters are then modified.

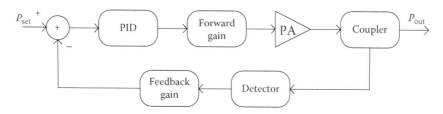

FIGURE 4.5   Typical closed-loop control for RF power amplifier for linearity control.

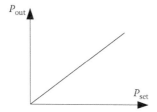

FIGURE 4.6   Linear curve for RF amplifier.

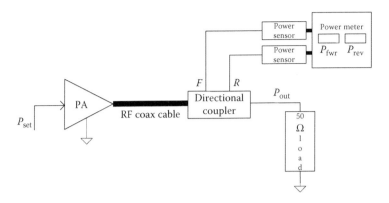

FIGURE 4.7   Experimental setup for linearity adjustment of RF power amplifiers.

4.1.5   1 dB COMPRESSION POINT

Compression point for an amplifier is the point where amplifier gain becomes 1 dB below its ideal linear gain (Figure 4.8). Once the 1 dB compression point is identified for the corresponding input power range, the amplifier can be operated in linear or nonlinear mode. Hence, the 1 dB compression point can also be conveniently used to identify the linear characteristics of the amplifier.

Gain at 1 dB compression point can be calculated as

$$P_{1dB,out} - P_{1dB,in} = G_{1dB} = G_0 - 1 \tag{4.11}$$

where G_0 is the small signal or linear gain of the amplifier at fundamental frequency. The 1 dB compression point can also be expressed using the input and output

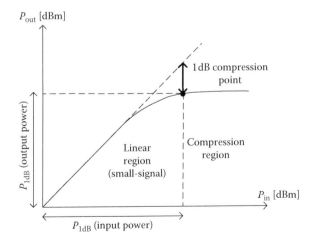

FIGURE 4.8   1 dB compression point for amplifiers.

voltages and their coefficients. The gain of the amplifier at fundamental frequency when $v_i(t) = \beta \cos \omega t$ is calculated as

$$G_{1dB} = 20 \log \left| \alpha_1 + 3\alpha_3 \frac{\beta^2}{4} \right| \tag{4.12}$$

$$G_0 \left(\text{Linear/small signal gain} \right) = 20 \log |\alpha_1| \tag{4.13}$$

As a result, the 1 dB compression point in Figure 4.8 can be calculated as

$$20 \log \left| \alpha_1 + 3\alpha_3 \frac{\beta_{in,1dB}^2}{4} \right| = 20 \log |\alpha_1| - 1 \, dB \tag{4.14}$$

where β_{1dB} is the amplitude of the input voltage at 1 dB compression point. Solution of Equation 4.14 for β_{1dB} leads to

$$\beta_{1dB} = \sqrt{0.145 \left| \frac{\alpha_1}{\alpha_3} \right|} \tag{4.15}$$

4.1.6 HARMONIC DISTORTION

Harmonic distortion for an amplifier can be defined as the ratio of the amplitude of the $n\omega$ component to the amplitude of the fundamental component. The second-order and third-order harmonic distortions can then be expressed as

$$HD_2 = \frac{1}{2} \frac{\alpha_2}{\alpha_1} \beta \tag{4.16}$$

$$HD_3 = \frac{1}{4} \frac{\alpha_3}{\alpha_1} \beta^2 \tag{4.17}$$

From Equations 4.16 and 4.17, it is apparent that the second harmonic distortion is proportional to signal amplitude, whereas third-order amplitude is proportional to the square of the amplitude. Hence, when input signal is increased by 1 dB, HD_2 increases by 1 dB and HD_3 increases by 2 dB. The total harmonic distortion (THD) in the amplifier can be calculated as

$$THD = \sqrt{\left(HD_2 \right)^2 + \left(HD_3 \right)^2 + \cdots} \tag{4.18}$$

Example 4.3

RF signal, $v_i(t) = \beta \cos \omega t$, is applied to a linear amplifier and then to a nonlinear amplifier given in Figure 4.9 with output response $v_o(t) = \alpha_0 + \alpha_1 \beta \cos \omega t + \alpha_2 \beta^2 \cos^2 \omega t + \alpha_3 \beta^3 \cos^3 \omega t$. Assume input and output impedances are equal to R. (1) Calculate

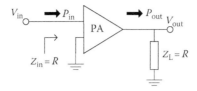

FIGURE 4.9 PA output response.

and plot gain for linear amplifier. (2) Obtain second and third harmonic distortions for nonlinear amplifier when $\alpha_0 = 0, \alpha_1 = 1, \alpha_2 = 3, \alpha_3 = 1$ and $\beta = 1, \beta = 2$. Calculate also THD for both cases.

Solution:

    1. For linear amplifier characteristics, the output voltage is expressed as

$$v_o(t) = \beta v_i(t) \tag{4.19}$$

which can also be written as

$$\frac{1}{2R} v_o^2(t) = \beta^2 \frac{1}{2R} v_i^2(t) \tag{4.20}$$

or

$$P_o = \beta^2 P_i \tag{4.21}$$

When power gain calculated from Equation 4.13 is given in dBm, then this can be expressed as

$$10\log\left(\frac{P_o}{1\text{ mW}}\right) = 10\log\left(\beta^2 \frac{P_i}{1\text{ mW}}\right) \tag{4.22}$$

or

$$P_o[\text{dBm}] = 10\log\left(\beta^2\right) + P_{in}[\text{dBm}] \tag{4.23}$$

Then, the power gain is obtained from Equation 4.14 as

$$\text{Gain}[\text{dBm}] = G[\text{dBm}] = 10\log\left(\beta^2\right) = P_o[\text{dBm}] - P_{in}[\text{dBm}] \tag{4.24}$$

The relation between input and output power is plotted and illustrated in Figure 4.10.

    2. The nonlinearity response of the amplifier using third-order polynomial can be expressed using Equation 1.2 as

$$v_o(t) = \alpha_0 + \alpha_1 v_i(t) + \alpha_2 v_i^2(t) + \alpha_3 v_i^3(t) \tag{4.25a}$$

or

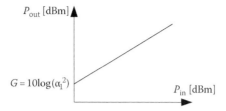

FIGURE 4.10 Power gain for linear operation.

$$v_o(t) = \alpha_0 + \alpha_1 \beta \cos \omega t + \alpha_2 \beta^2 \cos^2 \omega t + \alpha_3 \beta^3 \cos^3 \omega t \qquad (4.25b)$$

As seen from Equation 4.25, we have fundamental, second-order harmonic, third-order harmonic, and a DC component in the output response of the amplifier. In Equation 4.25, it is also seen that the DC component exists due to second harmonic content. Equation 4.25 can be rearranged to give the following closed form relation:

$$v_o(t) = \alpha_0 + \alpha_1 \beta \cos \omega t + \frac{\alpha_2 \beta^2}{2} + \frac{\alpha_2 \beta^2}{2} \cos 2\omega t + 3 \frac{\alpha_3 \beta^3}{4} \cos \omega t + \frac{\alpha_3 \beta^3}{4} \cos 3\omega t$$

$$(4.26a)$$

which can be simplified to

$$v_o(t) = \left(\alpha_0 + \frac{\alpha_2 \beta^2}{2} \right) + \left(\alpha_1 + 3 \frac{\alpha_3 \beta^2}{4} \right) \beta \cos \omega t + \left(\frac{\alpha_2}{2} \right) \beta^2 \cos 2\omega t + \frac{\alpha_3 \beta^3}{4} \cos 3\omega t$$

$$(4.26b)$$

When $\alpha_1 = 1$, $\alpha_2 = 3$, $\alpha_3 = 1$, and $\beta = 1$, HD_2 and HD_3 are calculated from Equations 4.16 and 4.17 as

$$HD_2 = \frac{1}{2} \frac{\alpha_2}{\alpha_1} \beta = \frac{1}{2} \frac{3}{1} (1) = 1.5 \qquad (4.27)$$

$$HD_3 = \frac{1}{4} \frac{\alpha_3}{\alpha_1} \beta^2 = \frac{1}{4} \frac{1}{1} (1)^2 = 0.25 \qquad (4.28)$$

When $\alpha_1 = 1$, $\alpha_2 = 3$, $\alpha_3 = 1$, and $\beta = 2$,

$$HD_2 = \frac{1}{2} \frac{\alpha_2}{\alpha_1} \beta = \frac{1}{2} \frac{3}{1} (2) = 3 \qquad (4.29)$$

$$HD_3 = \frac{1}{4} \frac{\alpha_3}{\alpha_1} \beta^2 = \frac{1}{4} \frac{1}{1} (2)^2 = 1 \qquad (4.30)$$

The THD for this system calculated from Equation 4.18 is

$$THD = \sqrt{(1.5)^2 + (0.25)^2} = 1.5625 \qquad (4.31a)$$

and

$$\text{THD} = \sqrt{(3)^2 + (1)^2} = 3.16 \qquad (4.31b)$$

Example 4.4

Input voltage given for an RF circuit is $v_{in}(t) = \beta \cos(\omega t)$. RF circuit generates signal at third harmonic as $V_3 \cos(3\omega t)$. What is the 1 dB compression point?

Solution: Using Equation 4.26b, the amplitude of the third harmonic component can be found from

$$\frac{\alpha_3 \beta^3}{4} = V_3 \text{ or } \alpha_3 = \frac{4V_3}{\beta^3} \qquad (4.32a)$$

Then, 1 dB compression point, β_{1dB}, is found from Equation 4.15 as

$$\beta_{1dB} = \sqrt{0.145 \left| \frac{\alpha_1}{\alpha_3} \right|} = \sqrt{\frac{0.145}{4} \left| \frac{\beta^3 \alpha_1}{V_3} \right|} = 0.19 \sqrt{\left| \frac{\beta^3 \alpha_1}{V_3} \right|} \qquad (4.32b)$$

4.1.7 INTERMODULATION

When a signal comprising two cosine waveforms with different frequencies

$$v_i(t) = \beta_1 \cos \omega_1 t + \beta_2 \cos \omega_2 t \qquad (4.33)$$

is applied to an input of an amplifier, the output signal consists of components of the self-frequencies and their products created by frequencies by ω_1 and ω_2 given by the equation below.

$$v_o(t) = \alpha_1 \left(\beta_1 \cos \omega_1 t + \beta_2 \cos \omega_2 t \right) + \alpha_2 \left(\beta_1 \cos \omega_1 t + \beta_2 \cos \omega_2 t \right)^2$$
$$+ \alpha_3 \left(\beta_1 \cos \omega_1 t + \beta_2 \cos \omega_2 t \right)^3 \qquad (4.34a)$$

or

$$v_o(t) = \alpha_1 \left(\beta_1 \cos \omega_1 t + \beta_2 \cos \omega_2 t \right) + \alpha_2 \left(\begin{array}{c} \frac{1}{2}\beta_1^2 \left(1 + \cos 2\omega_1 t\right) + \frac{1}{2}\beta_2^2 \left(1 + \cos 2\omega_2 t\right) + \\ \frac{1}{2}\beta_1\beta_2 \left(\cos(\omega_1 + \omega_2)t + \cos(\omega_1 - \omega_2)t \right) \end{array} \right)$$

$$+ \alpha_3 \left(\begin{array}{c} \frac{3}{4}\beta_1^3 \left(\cos \omega_1 t\right) + \frac{3}{2}\beta_1\beta_2^2 \left(\cos \omega_1 t\right) + \frac{1}{4}\beta_1^3 \cos(3\omega_1 t) + \frac{3}{4}\beta_1\beta_2^2 \left(\cos(\omega_1 - 2\omega_2)t\right) + \\ \frac{3}{4}\beta_1^2\beta_2 \left(\cos(2\omega_1 - \omega_2)t\right) + \frac{3}{2}\beta_1^2\beta_2 \left(\cos \omega_2 t\right) + \frac{3}{4}\beta_2^3 \left(\cos \omega_2 t\right) + \frac{1}{4}\beta_2^3 \left(\cos 3\omega_2 t\right) + \\ \frac{3}{4}\beta_1^2\beta_2 \left(\cos(2\omega_1 + \omega_2)t\right) + \frac{3}{4}\beta_1\beta_2^2 \left(\cos(\omega_1 + 2\omega_2)t\right) \end{array} \right)$$

$$(4.34b)$$

In Equation 4.34, DC component, α_0, is ignored. The components that will rise due to combinations of the frequencies, ω_1 and ω_2, as given by Equation 4.34a are shown in Table. 4.1. The corresponding frequency components in Table 4.1 are also illustrated in Figure 4.11.

In amplifier applications, intermodulation distortion products are undesirable components in the output signal. As a result, the amplifier needs to be tested using an input signal, which is the sum of two cosines, to eliminate these side products. This test is also known as a *two-tone test*. This specific test is important for an amplifier specifically when two frequencies, ω_1 and ω_2, are close to each other.

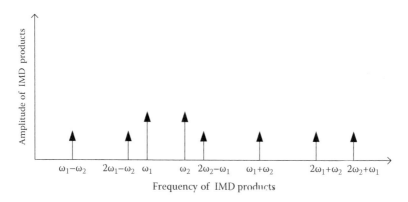

FIGURE 4.11 Illustration of IMD frequencies and products.

TABLE 4.1

Intermodulation Frequencies and Corresponding Amplitudes

$\omega = \omega_1$	$\left(\alpha_1\beta_1 + \dfrac{3}{4}\alpha_3\beta_1^3 + \dfrac{3}{2}\alpha_3\beta_1\beta_2^2\right)\cos(\omega_1 t)$
$\omega = \omega_2$	$\left(\alpha_1\beta_2 + \dfrac{3}{4}\alpha_3\beta_2^3 + \dfrac{3}{2}\alpha_3\beta_2\beta_1^2\right)\cos(\omega_2 t)$
$\omega = \omega_1 + \omega_2$	$\dfrac{1}{2}(\alpha_2\beta_1\beta_2)\cos(\omega_1 + \omega_2)t$
$\omega = \omega_1 - \omega_2$	$\dfrac{1}{2}(\alpha_2\beta_1\beta_2)\cos(\omega_1 - \omega_2)t$
$\omega = 2\omega_1 + \omega_2$	$\left(\dfrac{3}{4}\alpha_3\beta_1^2\beta_2\right)\cos(2\omega_1 + \omega_2)t$
$\omega = 2\omega_1 - \omega_2$	$\left(\dfrac{3}{4}\alpha_3\beta_1^2\beta_2\right)\cos(2\omega_1 - \omega_2)t$
$\omega = 2\omega_2 + \omega_1$	$\left(\dfrac{3}{4}\alpha_3\beta_1^2\beta_2\right)\cos(2\omega_2 + \omega_1)t$
$\omega = 2\omega_2 - \omega_1$	$\left(\dfrac{3}{4}\alpha_3\beta_1^2\beta_2\right)\cos(2\omega_2 - \omega_1)t$

The second-order intermodulation distortion, IM_2, can be found from Equation 4.18 and Table 4.1, when $\beta_1 = \beta_2 = \beta$. It is the ratio of the components at $\omega_1 \pm \omega_2$ to the fundamental components at ω_1 or ω_2.

$$IM_2 = \frac{\alpha_2}{\alpha_1}\beta \qquad (4.35)$$

The third-order distortion, IM_3, can be found from the ratio of the component at $2\omega_2 \pm \omega_1$
(or $2\omega_1 \pm \omega_2$) to the fundamental components at ω_1 or ω_2.

$$IM_3 = \frac{3}{4}\frac{\alpha_3}{\alpha_1}\beta^2 \qquad (4.36)$$

IM product frequencies are summarized in Table 4.2.

If Equations 4.16 through 4.17 and 4.35 through 4.36 are compared, intermodulation (IM) products can be related to harmonic distortion (HD) products as

$$IM_2 = 2HD_2 \qquad (4.37)$$

$$IM_3 = 3HD_3 \qquad (4.38)$$

IM_3 distortion components at frequencies $2\omega_1 - \omega_2$ and $2\omega_2 - \omega_1$ are very close to the fundamental components. That is the reason why IM_3 signal is measured most of the time for intermodulation distortion (IMD) characterization of the amplifier. The simplified measurement setup for IMD testing is shown in Figure 4.12.

The point at which the output components at the fundamental frequency and IM_3 intersect is called an intercept point or IP_3. At this point, $IM_3 = 1$, and IP_3 is found from Equation 4.36 as

---

TABLE 4.2

Summary of IM Product Frequencies

IM_2 frequencies	$\omega_1 \pm \omega_2$	
IM_3 frequencies	$2\omega_1 \pm \omega_2$	$2\omega_2 \pm \omega_1$
IM_5 frequencies	$3\omega_1 \pm 2\omega_2$	$3\omega_2 \pm 2\omega_1$

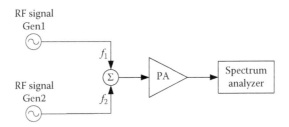

FIGURE 4.12   Simplified IMD measurement setup.

$$IM_3 = 1 = \frac{3}{4}\frac{\alpha_3}{\alpha_1}\left(IP_3\right)^2 \tag{4.39}$$

or

$$IP_3 = \sqrt{\frac{4}{3}\frac{\alpha_1}{\alpha_3}} \tag{4.40}$$

which can also be written as

$$IP_3 = \frac{V_{in}}{\sqrt{IM_3}} \tag{4.41}$$

where V_{in} is the input voltage. Equation 4.41 can be expressed in terms of dB by taking log of both sides in Equation 4.41 as

$$IP_3[dB] = V_{in}[dB] - \frac{1}{2}IM_3[dB] \tag{4.42}$$

The dynamic range (DR) is measured to understand the level of the output noise and is defined as

$$DR = \alpha_1 \frac{V_{in}}{V_{Nout}} = \frac{V_{in}}{V_{Nin}} \tag{4.43}$$

where input noise is related to output noise by

$$V_{Nin} = \frac{V_{Nout}}{\alpha_1} \tag{4.44}$$

So,

$$DR[dB] = V_{in}[dB] - V_{Nin}[dB] \tag{4.45}$$

Intermodulation-free dynamic range (IMFDR$_3$) is defined as the largest DR possible with no IM$_3$ product. For the third-order intermodulation distortion, V_{Nout} is defined as

$$V_{Nout} = \frac{3}{4}\alpha_3 V_{in}^3 \tag{4.46}$$

We can then obtain

$$V_{in} = \sqrt[3]{\frac{4}{3\alpha_3}V_{Nout}} \tag{4.47}$$

Substituting Equation 4.46 into Equation 4.43 gives IMFDR$_3$ as

$$DR = IMFDR_3 = \alpha_1 \frac{V_{in}}{V_{Nout}} = V_{in} = \sqrt[3]{\frac{4}{3}\frac{\alpha_1^3}{\alpha_3}\frac{1}{V_{Nout}^2}} \tag{4.48}$$

Since $V_{Nout} = \alpha_1 V_{Nin}$ from Equation 4.41, Equation 4.43 can be written in terms of input noise as

$$\text{IMFDR}_3 = \alpha_1 \frac{V_{in}}{V_{Nout}} = V_{in} = \sqrt[3]{\frac{4}{3}\frac{\alpha_1}{\alpha_3}\frac{1}{V_{Nin}^2}} \tag{4.49}$$

When Equations 4.40 and 4.49 are compared, IMFDR_3 can also be expressed using intercept point, IP_3, as

$$\text{IMFDR}_3 = \left(\frac{\text{IP}_3}{V_{Nin}}\right)^{2/3} \tag{4.50}$$

or in terms of dB, Equation 4.50 can also be given as

$$\text{IMFDR}_3[\text{dB}] = \frac{2}{3}\left(\text{IP}_3[\text{dB}] - V_{Nin}[\text{dB}]\right) \tag{4.51}$$

The relationship between the fundamental component and the third-order distortion component via input and output voltages is illustrated in Figure 4.13. In Figure 4.13, −1 dB compression point is used to characterize IM_3 product. The −1 dB compression point can be defined as the value of the input voltage, V_{in}, which is designated as $V_{in,1dBc}$, where fundamental component is reduced by 1 dB. $V_{in,1dBc}$ can be defined as

$$V_{in,1dBc} = \sqrt{(0.122)\left(\frac{4}{3}\right)\left|\frac{\alpha_1}{\alpha_3}\right|} \tag{4.52}$$

which is also equal to

$$V_{in,1dBc} = \sqrt{(0.122)}\text{IP}_3 \tag{4.53}$$

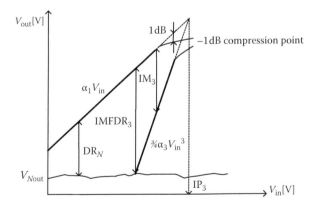

FIGURE 4.13   Illustration of the relation between fundamental components and IM_3.

Equation 4.53 can be expressed in [dB] as

$$V_{in,1dBc}(dB) = IP_3[dB] - 9.64\, dB \qquad (4.54)$$

As a result, once IP_3 is determined, Equation 4.53 can be used to calculate $-1\, dB$ compression point for the amplifier.

Example 4.5

Assume a sinusoid signal, $v_i(t) = \sin(\omega t)$, with $5\,Hz$ frequency is applied to a nonlinear amplifier that has output signal (1) $v_o(t) = 10\sin(\omega t) + 2\sin^2(\omega t)$, (2) $v_o(t) = 10\sin(\omega t) - 3\sin^3(\omega t)$, and (3) $v_o(t) = 10\sin(\omega t) + 2\sin^2(\omega t) - 3\sin^3(\omega t)$. Obtain the time domain representation of the input signal and frequency domain representation of power spectra of the output signal of the amplifier.

Solution:

1. The frequency spectrum for the input signal and power spectrum for the output signal of the amplifier are obtained using MATLAB® script given later. Based on the results shown in Figure 4.14, the amplifier output has components at DC, f, and $2f$.
2. The third-order response is shown in Figure 4.15. As seen from Figure 4.15, output signal does not have DC component anymore. The third-order effect shows itself as clipping in the time domain signal and fundamental and third-order components at the output power spectra of the signal.
3. Using the modified MATLAB script in parts (a) and (b), the time domain and frequency domain signals are obtained and illustrated in Figure 4.16.

As illustrated, the output response has components at DC, f, $2f$, and $3f$. Overall, the level of the nonlinearity response of the amplifier strongly depends on the coefficients of the output signal.

4.2 SMALL-SIGNAL AMPLIFIER DESIGN

There are three critical design stages that need to be implemented in small-signal amplifier design.

- Design of DC biasing circuit
- Obtaining parameters of the transistor
- Implementation of the small-signal amplifier design

The small-signal amplifier design begins with designing biasing circuit to choose the operating point, Q point, V_{ceq} and I_{cq} for BJT, and V_{dsq} and I_{dq} for field effect transistor (FET). In general, V_{ceq} and V_{dsq} are usually taken at half the system supply voltage for small-signal amplifier design [1–3]. Hence, once Q point is selected, designer is required to determine I_{cq} and I_{dq}.

After operating point is determined via DC biasing circuit, S parameters must be obtained at the operating Q point and the desired operating frequency, because

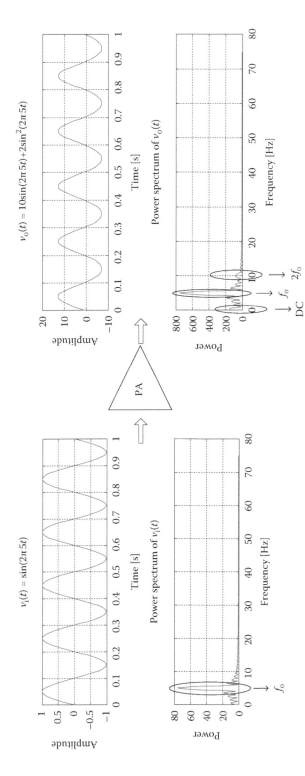

FIGURE 4.14 Second-order nonlinear amplifier output response that has components at DC, f, and $2f$.

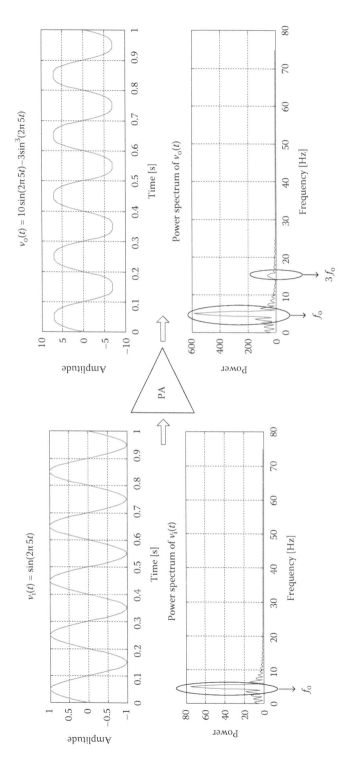

FIGURE 4.15 Third-order nonlinear amplifier output response that has components at f and $3f$.

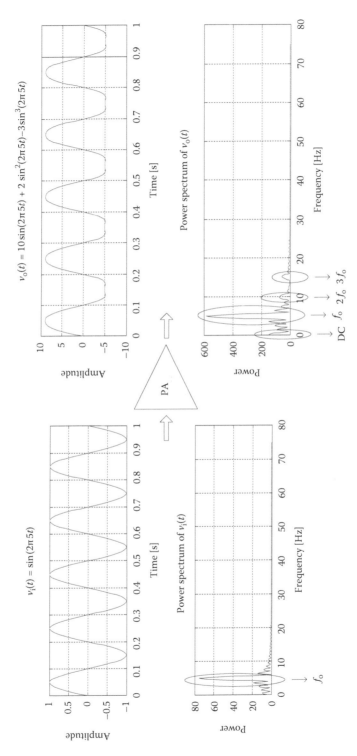

FIGURE 4.16   Nonlinear amplifier response that has second and third-order nonlinearities.

S parameters vary significantly based on the operating conditions. *S* parameters are provided by the manufacturer of the transistor most of the time. When DC biasing circuit is designed and *S* parameters are obtained, small-signal amplifier is then designed following the design stages that will be outlined in Section 4.2.4.

4.2.1 DC Biasing Circuits

Let us consider the BJT circuit given in Figure 4.17. When there is only DC source in the circuit, the i_c and v_{CE} curve can be obtained as shown in Figure 4.18.

If the circuit shown in Figure 4.17 is revised and AC sinusoidal source with amplitude ΔV_{BB} is introduced as shown in Figure 4.19, the base current and collector current will change to $i_B + \Delta i_B \cos(\omega t)$ and $i_C + \Delta i_C \cos(\omega t)$, respectively. Similarly, the collector-to-emitter voltage will change as $v_{CE} + \Delta v_{CE} \cos(\omega t)$.

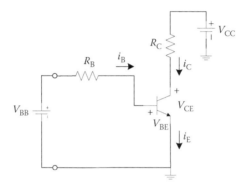

FIGURE 4.17 BJT circuit with only DC source.

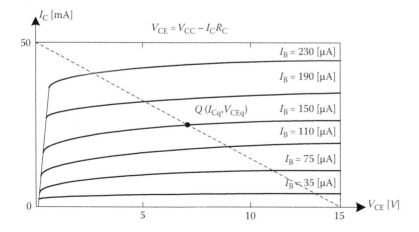

FIGURE 4.18 i_c and v_{CE} curve for DC bias.

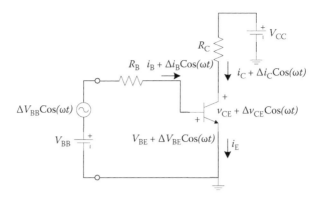

FIGURE 4.19 BJT circuit with DC and AC sources.

Consider the operational conditions for the BJT circuit under DC excitation as shown in Figure 4.18. Based on the given information, $I_B = 150$ [µA], $I_C = 23$ [mA], and $V_{CE} = 7.5$ [V] at Q point. When AC source is applied with DC excitation, base current, collector, and collector-to-emitter voltage change to

$$i_B + \Delta i_B = 150 + 40\cos(\omega t)\,[\text{mA}]$$

$$i_c + \Delta i_c = 22 + 7\cos(\omega t)\,[\text{mA}]$$

$$v_{CE} + \Delta v_{CE} = 7.5 - 2.5\cos(\omega t)\,[\text{V}]$$

This indicates that the base current will swing between 110 and 190 [µA]. One important criterion is to keep the operating points on the load line during swing of the current or voltage. When base current reaches its peak value to 190 [µA], collector-to-emitter voltage, V_{CE}, becomes around 4.5 [V], and collector current is equal to 29 [mA]. These changes can be illustrated in Figure 4.20.

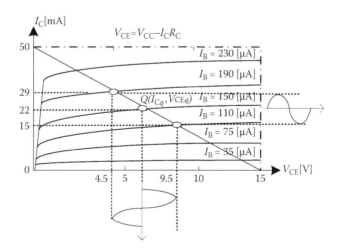

FIGURE 4.20 i_c and v_{CE} curve with DC and AC sources.

As seen from Figure 4.20, signal is amplified with the implementation of AC source. This is valid when the transistor, BJT, operates in the active region. If base current is increased further, the swing for the current will reach the value that will take transistor into the cutoff region.

4.2.2 BJT Biasing Circuits

There are typical biasing circuits that are used for amplifiers with BJTs. These biasing circuits can be referred to as (1) fixed bias, (2) stable bias, (3) emitter bias, and (4) self-bias.

4.2.2.1 Fixed Bias

Consider the common emitter configuration for the amplifier circuit given in Figure 4.21. Coupling capacitors are used to isolate the circuit from other circuits that are connected at the output and input of the amplifier. The common practice for this type of biasing circuit is to have V_{BB} and V_C equal to each other, so that it can be supplied from the single source. The idea of using the simple biasing circuit is to set a constant base current. However, this makes it very sensitive to gain variations, which then can be improved to some degree with R_1 and R_2. If BJT is assumed to be operating in the active linear region and $V_{BB} = V_C$, the circuit can be analyzed as follows: In our analysis, let us assume $R_1 \rightarrow 0$. So,

$$V_C = I_B R_2 + V_{BE} \tag{4.55}$$

which leads to

$$I_B = \frac{V_C - V_{BE}}{R_2} \tag{4.56}$$

So,

$$I_C = \beta I_B = \beta \frac{V_C - V_{BE}}{R_2} \tag{4.57}$$

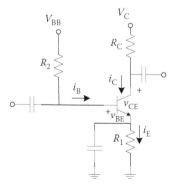

FIGURE 4.21 Fixed bias BJT circuit.

Collector-to-emitter voltage can be from

$$V_C = I_C R_C + V_{CE} \tag{4.58}$$

which leads to

$$V_{CE} = V_C - I_C R_C \tag{4.59}$$

When Equation 4.57 is substituted into Equation 4.59, we obtain

$$V_{CE} = V_C - R_C \left(\beta \frac{V_C - V_{BE}}{R_2} \right) \tag{4.60}$$

Example 4.6

Calculate R_1 and R_C for the fixed biasing circuit given in Figure 4.21, when $\beta = 100$ and $V_C = 15$ [V] to have the Q point at $V_{CE} = 7.5$ [V] and $I_C = 25$ [mA].

Solution: From Equation 4.55,

$$R_2 = \frac{V_C - V_{BE}}{I_B} = \frac{15 - 0.7}{0.25 \times 10^{-3}} = 57.2 \,[k\Omega]$$

R_C can be calculated from Equation 4.59 as

$$R_C = \frac{V_C - V_{CE}}{I_C} = \frac{15 - 7.5}{25 \times 10^{-3}} = 300 \,[\Omega]$$

4.2.2.2 Stable Bias

The stable bias circuit is different from the fixed bias circuit with the addition of the resistor between base and ground. R_3 can be adjusted with R_2 so that large improvement over stability can be achieved. Thevenin's theorem can be used to simplify the circuit shown in Figure 4.22 as illustrated in Figure 4.23. Typically, R_2 is 2 to 10 times greater than R_3.

In Figure 4.23,

$$V_{TH} = \frac{R_3}{R_2 + R_3} V_{CC} \tag{4.61a}$$

and

$$R_B = \frac{R_2 R_3}{R_2 + R_3} \tag{4.61b}$$

The stability in the circuit shown in Figure 4.23 can be described as follows. If the temperature increases, this will cause β to increase and results in an increase in collector current, I_C, and emitter current, I_E, beyond the desired operational values. However, because V_{TH} and R_B are fixed and do not vary, base current, I_B, will reduce and lower collector current, I_C, back to its original desired value. Therefore,

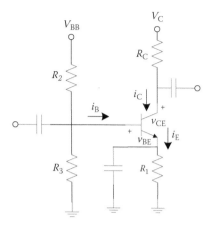

FIGURE 4.22 Stable bias circuit.

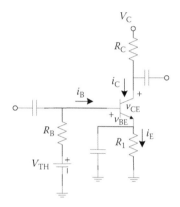

FIGURE 4.23 Simplified stable bias circuit.

the stability of the biasing network is satisfied with this method. We can assume in the circuit that

$$I_E \approx I_C = \beta I_B \tag{4.62}$$

In addition,

$$V_{TH} = I_B R_B + V_{BE} + I_E R_1 \tag{4.63}$$

So,

$$I_B = \frac{V_{TH} - V_{BE}}{R_B + \beta R_1} \tag{4.64}$$

and

$$V_C = I_C R_C + V_{CE} + I_E R_1 \qquad (4.65)$$

which leads to

$$V_{CE} = V_C - I_C (R_C + R_1) \qquad (4.66)$$

For feedback to be effective to have the stability,

$$R_B \ll \beta R_1 \qquad (4.67)$$

Hence,

$$I_B \approx \frac{V_{TH} - V_{BE}}{\beta R_1} \qquad (4.68)$$

So,

$$I_C \approx \frac{V_{TH} - V_{BE}}{R_1} \qquad (4.69)$$

and

$$V_{CE} = V_C - I_C (R_C + R_1) \approx V_C - \frac{R_C + R_1}{R_1} (V_{TH} - V_{BE}) \qquad (4.70)$$

It is clear from Equation 4.70 that I_C and V_{CE} do not depend on β anymore, which makes this circuit immune to the changes in β.

Example 4.7

Design a stable bias circuit shown in Figure 4.23 with a Q point of $I_C = 1.5$ [mA] and $V_{CE} = 7.5$ V when the DC gain of the transistor, β, ranges from 50 to 200.

Solution: We first find the collector supply voltage, V_C. Because we desire Q point located in the middle of the load line, then

$$V_C = 2V_{CE} = 2(7.5) = 15 \text{ [V]}$$

R_C and R_1 can be calculated from Equation 4.66 as

$$R_C + R_1 = \frac{V_C - V_{CE}}{I_C} = \frac{7.5}{1.5 \times 10^{-3}} = 5 \text{ [k}\Omega\text{]}$$

The designer can choose R_C and R_1 for the circuit. Let

$$R_C = 4 \text{ [k}\Omega\text{]} \text{ and } R_1 = 1 \text{ [k}\Omega\text{]}$$

We now need to find the base resistor, R_B, in Figure 4.23. It needs to satisfy the condition given in Equation 4.67. Then, we can get the lowest value of $\beta = 50$, and

$$R_B \ll \beta R_1 \rightarrow R_B = 0.1(50)(1000) = 5 \, [\text{k}\Omega]$$

From Equation 4.69,

$$V_{TH} \approx I_C R_1 + V_{BE} = 1.5 \times 10^{-3}(1 \times 10^{-3}) + 0.7 = 2.2 \, [\text{V}]$$

At this point, we can calculate R_3 and R_2 from Equation 4.61 as

$$\frac{V_{TH}}{V_C} = \frac{R_3}{R_2 + R_3} = 0.147 \text{ and } R_B = \frac{R_2 R_3}{R_2 + R_3} = 5000$$

Solution of the foregoing equation for R_3 and R_2 gives

$$R_2 = 5.8 R_3 \rightarrow R_2 = 24.734 \, [\text{k}\Omega] \text{ and } R_3 = 4.264 \, [\text{k}\Omega]$$

4.2.2.3 Self-Bias

In self-bias, R_C is connected through another resistor to the base of the circuit as feedback resistor. Because the voltage at the collector is lower than the fixed bias where it is directly supplied from V_C, the value of R_2 needs to be smaller. Let us analyze this circuit by assuming that BJT is operating in the linear active region, and $R_C \rightarrow 0$. Since $I_B \ll I_C$, then

$$I_1 = I_C + I_B \approx I_C \qquad (4.71)$$

Since

$$V_C = I_C R_C + V_{BE} + I_B R_2 \qquad (4.72a)$$

or

$$V_C = I_C \left(R_C + R_2 / \beta \right) + V_{BE} \qquad (4.72b)$$

and hence,

$$I_C = \frac{V_C - V_{BE}}{R_C + R_2 / \beta} \qquad (4.73)$$

If we assume $R_B / \beta \ll R_C$ and $V_{BE} = V\gamma$, then

$$I_C = \frac{V_C - V_\gamma}{R_C} \qquad (4.74)$$

Because the collector current does not depend on β, the bias point remains stable. The operational point in the active region can be checked to see whether $V_{CE} > V\gamma$ using

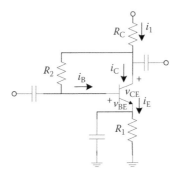

FIGURE 4.24 Self-bias circuit.

$$V_{CE} = R_2 I_B + V_{BE} = R_2 I_B + V_\gamma > V_\gamma \tag{4.75}$$

Equation 4.75 confirms that BJT is operating in the active region. The impact of the self-bias circuit can be understood better if we rewrite Equation 4.72a as

$$I_B = \frac{V_C - V_{BE} - I_C R_C}{R_2} = \frac{V_C - V_\gamma - I_C R_C}{R_2} \tag{4.76}$$

If β increases due to temperature change, the collector current will increase, and eventually, base current, I_B, will reduce. When I_B reduces, this will cause collector current, I_C, to drop. Hence, this will self-stable the change due to variation in β (Figure 4.24).

4.2.2.4 Emitter Bias

The most stable operating point can be achieved if the base of BJT circuit with common emitter configuration is grounded via R_2, and emitter is connected to another supply voltage as long as $R_2 \ll \beta R_1$. This can be verified by analyzing the circuit as

$$R_2 I_B + V_{BE} + R_1 I_E - V_E = 0 \tag{4.77}$$

Since

$$I_E \approx I_C = \beta I_B \tag{4.78}$$

then

$$R_2 \frac{I_E}{\beta} + R_1 I_E = V_E - V_{BE} \text{ or } I_E = \frac{V_E - V_{BE}}{R_1 + R_2 / \beta} \tag{4.79}$$

With the condition, such as $R_2 \ll \beta R_1$, we also obtain

$$I_C \approx I_E \approx \frac{V_E - V_{BE}}{R_1} = k_1 = \text{constant} \tag{4.80}$$

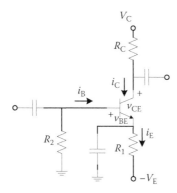

FIGURE 4.25 Emitter bias circuit.

So,

$$V_C = I_C R_C + V_{CE} + I_E R_1 - V_E \tag{4.81}$$

which leads to

$$V_{CE} = V_C + V_E - I_C (R_1 + R_C) = k_2 = \text{constant} \tag{4.82}$$

Hence, it is obvious from Equations 4.79, 4.80, and 4.82 that I_E, I_C, and V_{CE} do not depend on β, which shows having stable operation (Figure 4.25).

4.2.2.5 Active Bias Circuit

There are several other different biasing circuits that can be applied for BJTs and FETs. One of the methods to compensate for temperature variation in the biasing circuit is to use diodes.

For better temperature compensation, one of the methods used is implementation of diodes for compensation. The compensation of the temperature is achieved by the reduction of the internal resistance of diode against temperature increase. When temperature increases, diode's forward voltage reduces and causes base-to-emitter voltage to get lowered and compensate the effect of the increase in current due to temperature elevation. The biasing circuit using two diodes to accomplish this task is illustrated in Figure 4.26. The circuit shown in Figure 4.26a can be enhanced, and the number of components can be minimized by using active biasing controller IC such as BCR 400 by Siemens as shown in Figure 4.26b. This kind of biasing IC is able to supply stable bias current even at low supply voltage.

4.2.2.6 Bias Circuit Using Linear Regulator

If the bias voltage is desired to be independent of variations in the power supply, then bias circuit with linear regulator is a good option. In the sample bias circuit using linear regulator shown in Figure 4.27, temperature compensation is done using diode D1.

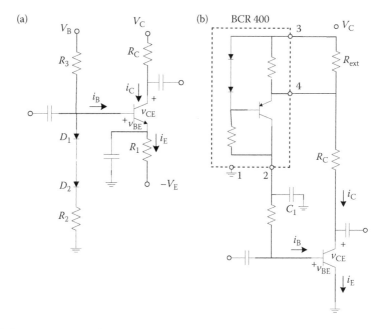

FIGURE 4.26 (a) Bias circuit with temperature-compensating diodes and (b) active bias circuit with BCR 400.

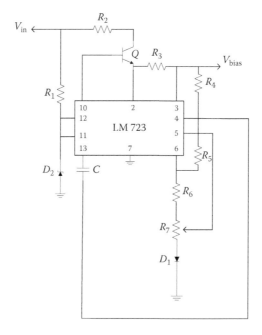

FIGURE 4.27 Bias circuit using linear regulator.

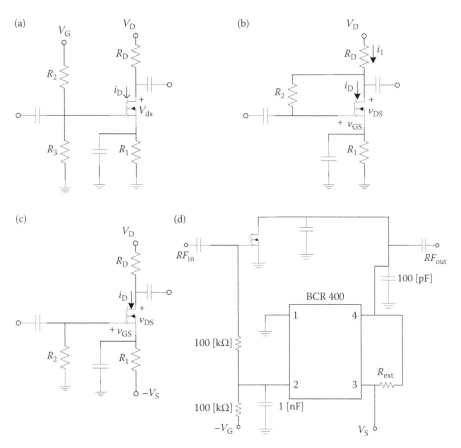

FIGURE 4.28 (a) Stable bias circuit, (b) self-bias circuit, (c) source bias circuit, and (d) active bias circuit.

4.2.3 FET Biasing Circuits

The biasing circuits for FETs can be constructed similar to circuits for BJTs. The bias circuit utilizing linear regulator shown in Figure 4.27 can also be applied for FETs. The important difference between FET and BJT circuits is that insignificantly small gate current flows with an application of input signal to gate; so the drain current can be assumed to be always equal to the source current. This enables FET circuits to be biased at their operating currents, which can be varied by the resistor connected to source, because it is always negative with respect to source voltage. The collection of similar biasing circuits for common source FETs are shown in Figure 4.28.

4.2.4 Small-Signal Amplifier Design Method

The design process begins with the selection of the operating points for the transistor as it was detailed in Section 4.2.3. Because the operating points for voltage and current are identified for the transistor, S parameters that are supplied by the manufacturer need to be obtained or measured. After S parameters are obtained or measured, it is now time to follow step-by-step design procedure to implement the small-signal amplifier. There are specific quantities and terms that will be used in the design of small-signal amplifiers. These terms and quantities will be explained in Section 4.2.4.1.

4.2.4.1 Definitions of Power Gains for Small-Signal Amplifiers

Consider the generalized two-port network shown in Figure 4.29. The illustration shown in Figure 4.30 in conjunction with Figure 4.29 can help to integrate the application of scattering parameters in amplifier design. In the design of small-signal amplifiers, three power quantities are used.

Let us obtain power relations for the circuit in Figure 4.29.

- At the input port,

$$V_1 = V_s \frac{Z_{in}}{Z_{in} + Z_s} = V_1^+ + V_1^- = V_1^+ \left(1 + \Gamma_{in}\right) \tag{4.83}$$

In addition,

$$\Gamma_{in} = \frac{Z_{in} - Z_o}{Z_{in} + Z_o} \tag{4.84}$$

$$\Gamma_s = \frac{Z_s - Z_o}{Z_s + Z_o} \tag{4.85}$$

$$\frac{Z_{in}}{Z_o} = \frac{1 + \Gamma_{in}}{1 - \Gamma_{in}} \tag{4.86}$$

So, from Equations 4.83 through 4.86, we can obtain the following relation:

$$V_1^+ = \frac{1}{1 + \Gamma_{in}} V_s \frac{Z_{in}}{Z_{in} + Z_s} \text{ or } V_1^+ = \frac{V_s\left(1 - \Gamma_s\right)}{2\left(1 - \Gamma_s\Gamma_{in}\right)} \tag{4.87}$$

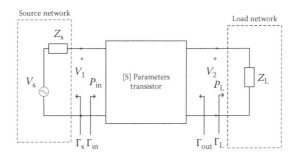

FIGURE 4.29 Generalized two-port network.

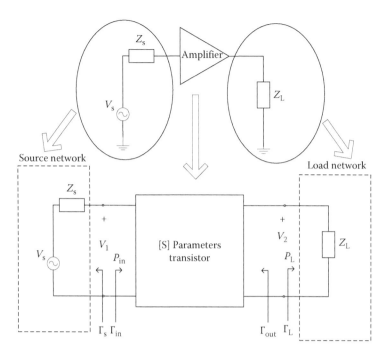

FIGURE 4.30 Integration of amplifier circuit.

As a result, input power can be written as

$$P_{\text{in}} = \frac{1}{2} \frac{\left|V_1^+\right|^2}{Z_{\text{o}}} \left(1 - \left|\Gamma_{\text{in}}\right|^2\right) \text{ or } P_{\text{in}} = \frac{1}{8} \frac{\left|V_s\right|^2}{Z_{\text{o}}} \frac{\left(1 - \left|\Gamma_{\text{in}}\right|^2\right)\left|1 - \Gamma_{\text{in}}\right|^2}{\left|1 - \Gamma_s \Gamma_{\text{in}}\right|^2} \tag{4.88}$$

Available power from source, P_{avs}, can be found when $\Gamma_{\text{in}} = \Gamma_s^*$ from

$$P_{\text{avs}}(\Gamma_s) = P_{\text{in}}\big|_{\Gamma_{\text{in}} = \Gamma_s^*} \tag{4.89a}$$

or

$$P_{\text{avs}}(\Gamma_s) = P_{\text{in}}\big|_{\Gamma_{\text{in}} = \Gamma_s^*} = \frac{1}{8} \frac{\left|V_s\right|^2}{Z_{\text{o}}} \frac{\left|1 - \Gamma_s\right|^2}{\left(1 - \left|\Gamma_s\right|^2\right)} \tag{4.89b}$$

- At the output port,

$$V_2^- = S_{21}V_1^+ + S_{22}V_2^+ \tag{4.90a}$$

$$V_2^+ = \Gamma_L V_2^- \tag{4.90b}$$

Substituting Equation 4.90b into Equation 4.90a gives

$$V_2^- = S_{21}V_1^+ + S_{22}\Gamma_L V_2^- \text{ or } V_2^- = \frac{S_{21}V_1^+}{1 - S_{22}\Gamma_L} = \frac{S_{21}V_s(1-\Gamma_s)}{(1-S_{22}\Gamma_L)(1-\Gamma_s\Gamma_{in})}$$

$$\text{or } V_2^- = \frac{S_{21}V_1^+}{1 - S_{22}\Gamma_L} = \frac{S_{21}V_s(1-\Gamma_s)}{(1-S_{22}\Gamma_L)(1-\Gamma_s\Gamma_{in})} \tag{4.91}$$

Then,

$$P_L = \frac{1}{2}\frac{|V_2^-|^2}{Z_o}\left(1-|\Gamma_L|^2\right) \text{ or } P_L = \frac{1}{8}\frac{|V_s|^2}{Z_o}\frac{|S_{21}|^2|1-\Gamma_s|^2}{|1-S_{22}\Gamma_L|^2(1-\Gamma_s\Gamma_{in})}\left(1-|\Gamma_L|^2\right) \tag{4.92}$$

Available power from network, P_{avn}, can be found when $\Gamma_L = \Gamma_{out}^*$ from

$$P_{avn}(\Gamma_{out}) = P_L\big|_{\Gamma_L=\Gamma_{out}^*} \tag{4.93a}$$

or

$$P_{avn}(\Gamma_{out}) = P_L\big|_{\Gamma_L=\Gamma_{out}^-} = \frac{1}{8}\frac{|V_s|^2}{Z_o}\frac{|S_{21}|^2|1-\Gamma_s|^2}{|1-S_{22}\Gamma_{out}^*|^2(1-\Gamma_s\Gamma_{in})}\left(1-|\Gamma_{out}|^2\right) \tag{4.93b}$$

Since,

$$\Gamma_{in} = S_{11} + \frac{S_{12}S_{21}\Gamma_L}{1-S_{22}\Gamma_L} = \frac{S_{11} - S_{11}S_{22}\Gamma_L + S_{12}S_{21}\Gamma_L}{1-S_{22}\Gamma_L} \tag{4.94}$$

Substituting Equation 4.94 into Equation 4.93b when $\Gamma_L = \Gamma_{out}^*$ gives

$$P_{avn}(\Gamma_{out}) = P_L\big|_{\Gamma_L=\Gamma_{out}^*} = \frac{1}{8}\frac{|V_s|^2}{Z_o}\frac{|S_{21}|^2|1-\Gamma_s|^2}{|1-S_{11}\Gamma_s|^2\left(1-|\Gamma_{out}|^2\right)^2} \tag{4.95}$$

There are three important power quantities that need to be known. They are operating power gain, G_p, available power gain, G_A, and transducer power gain, G_T.

- Transducer Gain

Transducer gain, G_T, is the ratio of the time-averaged power dissipated at the load and the maximally available time-averaged power from the source

$$G_T = \frac{P_L}{P_{avs}} \tag{4.96}$$

Substituting Equations 4.92 and 4.89b into Equation 4.96 gives

$$G_T = \frac{P_L}{P_{avs}} = \underbrace{\frac{\left(1-|\Gamma_s|^2\right)}{\left|1-\Gamma_s\Gamma_{in}\right|^2}}_{source} |S_{21}|^2 \underbrace{\frac{\left(1-|\Gamma_L|^2\right)}{\left|1-S_{22}\Gamma_L\right|^2}}_{load} \qquad (4.97)$$

The transducer power gain, G_T, is the gain component that is used to understand the amplifying level of the amplifier. If there is a specific power gain requirement for the design of an amplifier, G_T is the quantity that is always referred to.

• Operating Gain

Operating gain, G_p, is the ratio of the time-averaged power dissipated a the load to the time-averaged power delivered to network and obtained from

$$G_p = \frac{P_L}{P_{in}} \qquad (4.98)$$

Substituting Equations 4.88 and 4.92 into Equation 4.98 gives

$$G_p = \frac{P_L}{P_{in}} = \underbrace{\frac{1}{\left(1-|\Gamma_{in}|^2\right)}}_{Source} |S_{21}|^2 \underbrace{\frac{\left(1-|\Gamma_L|^2\right)}{\left|1-S_{22}\Gamma_L\right|^2}}_{Load} \qquad (4.99)$$

• Available Gain

Available gain, G_A, is the ratio of the maximally available time-averaged power from the network to the maximally available time-averaged power from the source and is defined by

$$G_A = \frac{P_{avn}}{P_{avs}} \qquad (4.100)$$

Substituting Equations 4.95 and 4.89b into Equation 4.100 gives

$$G_A = \frac{P_{avn}}{P_{avs}} = \underbrace{\frac{\left(1-|\Gamma_s|^2\right)}{\left|1-S_{11}\Gamma_s\right|^2}}_{source} |S_{21}|^2 \underbrace{\frac{1}{\left(1-|\Gamma_{out}|^2\right)}}_{load} \qquad (4.101)$$

Operating and available power gains, G_p and G_A, are used as knobs to meet the certain amplifier transducer gain, G_T.

Consider the more general and practical amplifier circuit shown in Figure 4.31. If S parameters of the transistor, Γ_s, and Γ_L of the circuit are known, then the gain components of the amplifier G_p, G_A, and G_T can be calculated using Equations 4.97, 4.99, and 4.101.

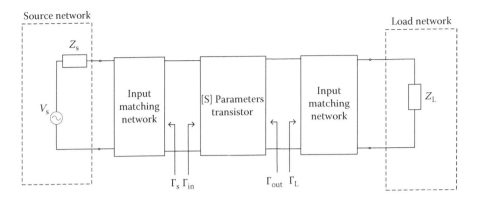

FIGURE 4.31   General two-port amplifier network.

The reflection coefficients that are used to calculate the gain parameters are

$$\Gamma_s = \frac{Z_s - Z_o}{Z_s + Z_o} \tag{4.102}$$

$$\Gamma_L = \frac{Z_L - Z_o}{Z_L + Z_o} \tag{4.103}$$

$$\Gamma_{in} = S_{11} + \frac{S_{12}S_{21}\Gamma_L}{1 - S_{22}\Gamma_L} \tag{4.104}$$

$$\Gamma_{out} = S_{22} + \frac{S_{12}S_{21}\Gamma_s}{1 - S_{11}\Gamma_s} \tag{4.105}$$

4.2.4.2   Design Steps for Small-Signal Amplifier

The following design steps, when followed, will simplify the design of small-signal amplifiers:

1. Design the biasing circuit based on transistor manufacturer datasheet
2. Obtain S parameters at bias conditions
3. Investigate the stability of the transistor. Check the stability using Rollet stability factor k. If $|k| > 1$ and $|\Delta| < 1$, then this implies that the transistor is unconditionally stable.
4. If the transistor is unconditionally stable, then design amplifier for gain.
5. If the condition mentioned in step 3 for unconditional stability is not met, then the transistor is potentially stable. If the transistor is potentially stable, draw source and load stability circles by calculating the center points and radii for input and output stability circles in Smith chart: r_{in}, C_{in}, and r_{out}, C_{out}.
6. Calculate whether Γ_L and Γ_s lie in an unstable region. If they lie in a stable region, then nothing more needs to be done for stability.

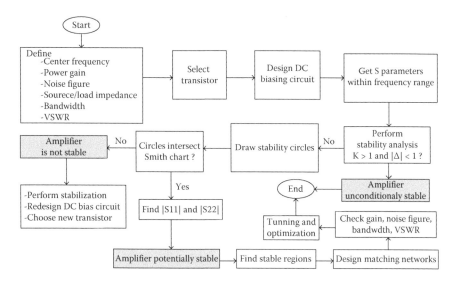

FIGURE 4.32   Small-signal amplifier design method illustration.

7. If Γ_L and Γ_s lie in an unstable region, then shunt or series resistances need to be added at the input and/or output to move S_{22}^* and/or S_{11}^* into stable region.
8. Adjust the gain of the amplifier if needed by Γ_L or Γ_s.
9. Plot the input and output gain circles for unilateral case if $S_{12} \to 0$. Use transducer power gain Equation 4.97. The specified gain must be less than the maximum unilateral gain, G_{TUmax}.
10. If $S_{12} \neq 0$, then the design of the amplifier using the unilateral method (to have control of selective mismatch at the input and output) can be done by investigating the maximum error based on the Unilateral Figure of Merit to determine whether the method could be applied.
11. If unilateral method cannot be applied, then the designer should proceed with the bilateral design method.

The aforementioned steps are simplified and illustrated in Figure 4.32.

4.2.4.3 Small-Signal Amplifier Stability

The stability analysis for amplifiers is necessary to measure the amplifier's resistance against oscillations. The oscillations may occur if reflected signals at the input or output port increase their magnitudes while they are reflected between an active port and its termination continuously. Oscillations in an amplifier are not desired because when they occur, the characteristic of the amplifier changes drastically. Scattering parameters become no longer valid, and hence circuit does not perform as expected. This may result in a catastrophic failure and may damage active device and surrounding components.

4.2.4.3.1 Unconditional Stability

When amplifier remains stable throughout the entire cycle under the operating conditions and frequency, the amplifier is said to be unconditionally stable. Unconditional stability can be satisfied for any passive source and load when

$$|\Gamma_s| < 1 \tag{4.106a}$$

and

$$|\Gamma_L| < 1 \tag{4.106b}$$

Having negative resistance can be avoided if

$$|\Gamma_s \Gamma_{in}| < 1 \tag{4.107a}$$

$$|\Gamma_L \Gamma_{out}| < 1 \tag{4.107b}$$

So,

$$|\Gamma_{in}| = \left| S_{11} + \frac{S_{12}S_{21}\Gamma_L}{1 - S_{22}\Gamma_L} \right| < 1 \tag{4.108}$$

$$|\Gamma_{out}| = \left| S_{22} + \frac{S_{12}S_{21}\Gamma_s}{1 - S_{11}\Gamma_s} \right| < 1 \tag{4.109}$$

A two-port network is said to be unconditionally stable at a given frequency if

$$\text{Re}\{Z_{in}\} > 0 \text{ and } \text{Re}\{Z_{out}\} > 0 \tag{4.110}$$

In practice, Roulette's Stability Factor, k, and determinant of the scattering matrix of the active device, $|\Delta|$, can also be used to test the unconditional stability of two-port networks using the following equations:

$$k = \frac{1 + |\Delta|^2 - |S_{11}|^2 - |S_{22}|^2}{2|S_{21}||S_{12}|} > 1 \tag{4.111}$$

$$|\Delta| = |S_{22}S_{11} - S_{12}S_{21}| < 1 \tag{4.112}$$

4.2.4.3.2 Stability Circles

We can investigate the stability of the amplifier graphically by studying the stability circles on the Smith chart.

4.2.4.3.2.1 Output Stability Circle
Γ_L or load plane on the Smith chart is defined by output stability circles where the boundaries are defined between $|\Gamma_L| < 1$ (stable) and $|\Gamma_L| < 1$ (unstable). This can be found by solving between Γ_s in the following equation:

$$|\Gamma_{in}| = \left| S_{11} + \frac{S_{12}S_{21}\Gamma_L}{1 - S_{22}\Gamma_L} \right| = 1 \tag{4.113}$$

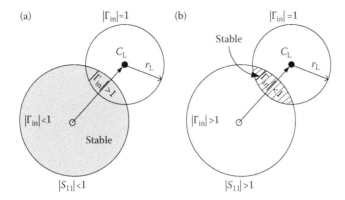

FIGURE 4.33   Smith chart illustrating output stability regions.

Solution of Equation 4.113 lies on a circle with radius

$$r_L = \left| \frac{S_{12}S_{21}}{|S_{22}|^2 - |\Delta|^2} \right| \tag{4.114}$$

center

$$c_L = \frac{\left(S_{22} - \Delta S_{11}^*\right)^*}{|S_{22}|^2 - |\Delta|^2} \tag{4.115}$$

where

$$|\Delta| = |S_{22}S_{11} - S_{12}S_{21}| \tag{4.116}$$

The circles formed by Equations 4.113 through 4.115 establish stability regions for the output. This can be done by plotting Γ_L circle and investigating the intersection of this circle with the Smith chart. If the circle intersects the Smith chart, then there is an instability region defined by the boundary. If there is no intersection, it means the device is unconditionally stable. The graphical illustration of the stability regions is given in Figure 4.33. The region inside the Smith chart, where $|S_{11}| < 1$, represents the stable region (represented by shaded regions in Figure 4.33).

4.2.4.3.2.2 Input Stability Circle Γ_s or source plane on the Smith chart is defined by input stability circles, where the boundaries are defined between $|\Gamma_s| < 1$ (stable) and $|\Gamma_s| < 1$ (unstable). This can be found by solving between Γ_L in the following equation:

$$|\Gamma_{out}| = \left| S_{22} + \frac{S_{12}S_{21}\Gamma_s}{1 - S_{11}\Gamma_s} \right| = 1 \tag{4.117}$$

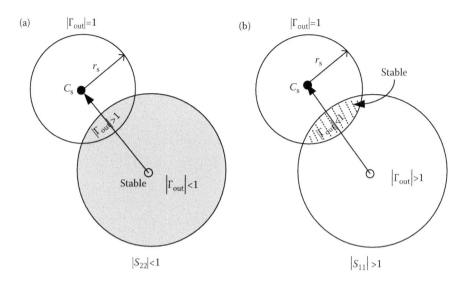

FIGURE 4.34   Smith chart illustrating input stability regions.

Solution of Equation 4.117 lies on a circle with radius and center as

$$r_s = \left| \frac{S_{12}S_{21}}{|S_{11}|^2 - |\Delta|^2} \right| \tag{4.118}$$

$$c_s = \frac{\left(S_{11} - \Delta S_{22}^* \right)^*}{|S_{11}|^2 - |\Delta|^2} \tag{4.119}$$

The circles formed by Equations 4.117 through 4.119 establish stability regions for the input. This can be done by plotting Γ_s circle and investigating the intersection of this circle with the Smith chart. The graphical illustration of the stability regions for input is given in Figure 4.34.

If the device is unconditionally stable, then there is no intersection of the circle with the Smith chart. This can be shown by Figure 4.35.

Example 4.8

S parameters of a transistor at 800 MHz are given to be $S_{11} = 0.68 < -72°$, $S_{12} = 0.18 < -14°$, $S_{21} = 4.5 < 82°$, and $S_{22} = 0.65 < -42°$. Determine the stability of the device, and draw the stability circles if the device is potentially stable.

Solution: Rollet's stability factor k and Δ are calculated from Equations 4.111 to 4.112 and found to be $k = 1.0381 > 1$. However, because $\Delta > 1.2509$, the device is still potentially unstable. The stability circles will then be inside the Smith chart. In addition, the scattering parameters $|S_{11}| < 1$ and $|S_{22}| < 1$ show that inside these

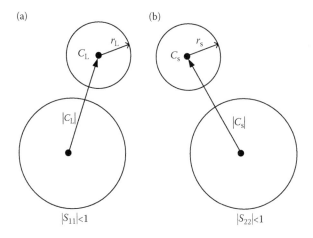

FIGURE 4.35 Unconditional stability (a) Γ_L plane and (b) Γ_s plane.

circles are stable, and the point $\Gamma_L = 0$ remains in the stable region. The parameters of stability circles from Equations 4.113 to 4.119 have to be calculated to plot them. The calculation of the radius and center point for input stability circle are

$$r_s = 0.7340 \text{ and } C_s = 0.5575 - j1.0801$$

The radius and center point for output stability circle are

$$r_L = 0.7083 \text{ and } C_L = -0.1927 - j1.0889$$

The results showing the stability circles and stable regions are shown in Figure 4.36.

4.2.4.3.3 Stabilization of Amplifier

In Figure 4.37, the stability in the circuit exits if

$$\left|\Gamma_{in}\right| = \left|S_{11} + \frac{S_{12}S_{21}\Gamma_L}{1 - S_{22}\Gamma_L}\right| < 1 \tag{4.120a}$$

$$\left|\Gamma_{out}\right| = \left|S_{22} + \frac{S_{12}S_{21}\Gamma_s}{1 - S_{11}\Gamma_s}\right| < 1 \tag{4.120b}$$

If $\left|\Gamma_{in}\right| > 1$ and $\left|\Gamma_{out}\right| > 1$, then this indicates the condition for instability. Γ_{in} and Γ_{out} can also be represented in terms of impedances as

$$\left|\Gamma_{in}\right| = \left|\frac{Z_{in} - Z_o}{Z_{in} + Z_o}\right| > 1 \tag{4.121a}$$

$$\left|\Gamma_{out}\right| = \left|\frac{Z_{out} - Z_o}{Z_{out} + Z_o}\right| > 1 \tag{4.121b}$$

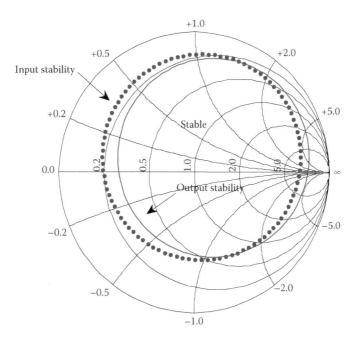

FIGURE 4.36   Stability circles for Example 4.8.

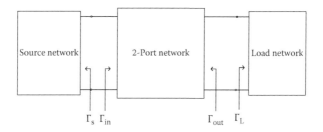

FIGURE 4.37   Two-port network for stabilization.

The Equation 4.121 imply that

$$\text{Re}\{Z_{\text{in}}\} < 0 \text{ or } \text{Re}\{Z_{\text{out}}\} < 0 \tag{4.122}$$

Reexpress Equation 4.121 as

$$\Gamma_{\text{in}} = \frac{(R_{\text{in}} - Z_{\text{o}}) + jX_{\text{in}}}{(R_{\text{in}} + Z_{\text{o}}) + jX_{\text{in}}} \rightarrow |\Gamma_{\text{in}}| = \sqrt{\frac{(R_{\text{in}} - Z_{\text{o}})^2 + X_{\text{in}}^2}{(R_{\text{in}} + Z_{\text{o}})^2 + X_{\text{in}}^2}} \tag{4.123}$$

Equation 4.123 shows that then the amplifier can be stabilized by adding series resistance or shunt conductance to source side to make the real part of the impedance positive. This can be similarly done for the load network, which gives

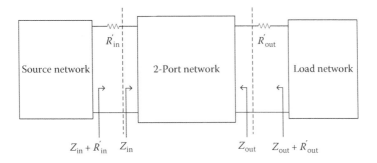

FIGURE 4.38 Stabilization network by adding series resistance.

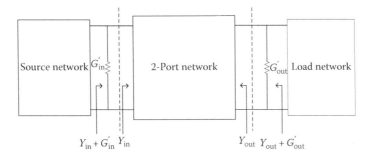

FIGURE 4.39 Stabilization network by adding shunt conductance.

$$\Gamma_{in} = \frac{(R_{out} - Z_o) + jX_{out}}{(R_{out} + Z_o) + jX_{out}} \rightarrow |\Gamma_{in}| = \sqrt{\frac{(R_{out} - Z_o)^2 + X_{out}^2}{(R_{out} + Z_o)^2 + X_{out}^2}} \qquad (4.124)$$

The addition of the stability components at the source and load is illustrated in Figures 4.38 and 4.39. This can be better understood by a simple illustrative example. Assume we have an amplifier with certain load impedance and load stability circle as shown in Figure 4.40. If we insert a series resistance of 15 Ω, we can then limit the stability region to Z', where $R = 15$ circle is tangential to the output stability circle in Figure 4.40. The stabilization can also be done by using shunt conductance (as shown in Figure 4.41).

Assume we have an amplifier with certain load impedance and load stability circle as shown in Figure 4.41. If we insert a shunt resistance of 400 Ω, we can then limit the stability region to Y', where $G = 0.0025$ circle is tangential to the output stability circle in Figure 4.41.

4.2.4.4 Constant Gain Circles

4.2.4.4.1 Unilateral Case

When $|S_{12}| \rightarrow 0$, the condition of the two-port network is called as unilateral. It is practical to obtain and plot the gain stability circles for unilateral devices. Under unilateral condition, Γ_{in} from Equation 4.104 becomes

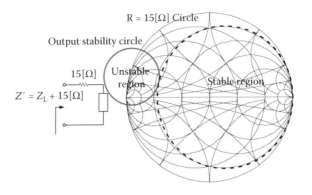

FIGURE 4.40   Stabilization with series resistor at the load.

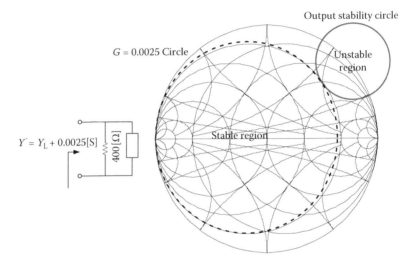

FIGURE 4.41   Stabilization with shunt conductance at the load.

$$\Gamma_{in} = S_{11} + \frac{S_{12}S_{21}\Gamma_L}{1 - S_{22}\Gamma_L} \rightarrow \Gamma_{in} = S_{11} \qquad (4.125)$$

Substituting Equation 4.125 in Equation 4.104 gives transducer gain of the unilateral device as

$$G_T = \underbrace{\frac{\left(1 - |\Gamma_s|^2\right)}{|1 - \Gamma_s S_{11}|^2}}_{G_s} \underbrace{|S_{21}|^2}_{G_o} \underbrace{\frac{\left(1 - |\Gamma_L|^2\right)}{|1 - S_{22}\Gamma_L|^2}}_{G_L} \qquad (4.126)$$

Equation 4.126 can be expressed as

$$G_T = G_s G_o G_L \qquad (4.127a)$$

Or

$$G_T[dB] = G_s[dB] + G_o[dB] + G_L[dB] \qquad (4.127b)$$

where

$$G_s = \frac{\left(1 - |\Gamma_s|^2\right)}{|1 - \Gamma_s S_{11}|^2} \qquad (4.128)$$

$$G_o = |S_{21}|^2 \qquad (4.129)$$

$$G_L = \frac{\left(1 - |\Gamma_L|^2\right)}{|1 - S_{22}\Gamma_L|^2} 1 \qquad (4.130)$$

G_s and G_L are the gain contributions for input and output matching networks, respectively. G_L is the gain of the transistor as shown in Figure 4.42. In terms of dB, Equation 4.127 can be written as

$$G_T[dB] = G_s[dB] + G_o[dB] + G_L[dB] \qquad (4.131)$$

The power transfer for the unilateral case can be maximized when both input and output ports of the amplifier are conjugately matched as:

$$\Gamma_s = S_{11}^* \qquad (4.132)$$

$$\Gamma_L = S_{22}^* \qquad (4.133)$$

When simultaneous conjugate match is accomplished, the maximum value of the gain is obtained, and the following relation holds:

$$G_{Tmax} = G_{pmax} = G_{Amax} \qquad (4.134)$$

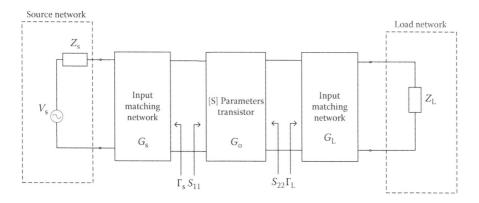

FIGURE 4.42   Unilateral amplifier design.

So,

$$G_{T,max} = G_{s,max}G_oG_{L,max} = \frac{1}{\left|1-S_{11}\right|^2}\left|S_{21}\right|^2\frac{1}{\left|1-S_{22}\right|^2} \tag{4.135}$$

where

$$G_{s,max} = \frac{1}{\left|1-S_{11}\right|^2} \tag{4.136}$$

$$G_{L,max} = \frac{1}{\left|1-S_{22}\right|^2} \tag{4.137}$$

Expressions for G_s and G_L in Equations 4.136 and 4.137 can be generalized as

$$G_i = \frac{1-\left|\Gamma_i\right|^2}{\left|1-S_{ii}\Gamma_i\right|^2} \tag{4.138a}$$

$$G_{i,max} = \frac{1}{\left|1-S_{ii}\right|^2} \tag{4.138b}$$

where $i = s$ or L and $i = 11$ or 22. Hence, the design for a specified amplifier gain can be done using Equation 4.138. There are two cases to consider under the unilateral condition for constant gain amplifier design: $|S_{ii}| < 0 \rightarrow$ Unconditionally Stable Case and $|S_{ii}| > 0 \rightarrow$ Potentially Stable Case.

4.2.4.4.2 $|S_{ii}| < 1 \rightarrow$ Unilateral Unconditionally Stable Case
The maximum gain for source and load shown in Equations 4.136 and 4.137 is obtained when simultaneous conjugate matching is done using

$$G_{i,max} = \frac{1}{\left|1-S_{ii}\right|^2} \tag{4.139}$$

$$\Gamma_i = S_{ii}^* \tag{4.140}$$

When $\left|\Gamma_i\right| = 1$, gain gets its lowest value from Equation 4.138a. The range of the gain value for values different from $\left|\Gamma_i\right| \neq 1$ is

$$0 \leq G_i \leq G_{i,max} \tag{4.141}$$

The constant gain circles lie inside the Smith chart and are obtained by solving the following equation:

$$\left|\Gamma_i - C_{gi}\right| = r_{gi} \tag{4.142}$$

In Equation 4.142, C_{gi} represents the center and r_{gi} is the radius of the constant gain circle, which are defined by

$$C_{gi} = \frac{g_i S_{ii}^*}{1 - |S_{ii}|^2 (1 - g_i)}$$ (4.143)

$$r_{gi} = \frac{\sqrt{1 - g_i} \left(1 - |S_{ii}|^2\right)}{1 - |S_{ii}|^2 (1 - g_i)}$$ (4.144)

where g_i is the normalized gain factor and expressed as

$$g_i = \frac{G_i}{G_{i,\max}} = \frac{1 - |\Gamma_i|^2}{|1 - S_{ii} \Gamma_i|^2} \left(1 - |S_{ii}|^2\right)$$ (4.145)

where

$$0 \le g_i \le 1$$ (4.146)

The gain circles are obtained by the following method:

- Plot S_{ii}^* on the Smith chart. Draw a line from the center of the Smith chart to S_{ii}^*. This is the point where gain is maximum.
- Calculate G_i and normalized g_i from Equations 4.138a and 4.145, respectively.
- Calculate center and radius points for corresponding G_i using Equations 4.143 and 4.144, and draw the constant gain circles.

The process of drawing gain circles can be illustrated in Figure 4.43.

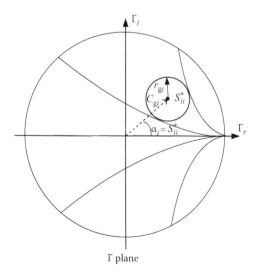

FIGURE 4.43 Drawing constant gain circles.

4.2.4.4.3 $|S_{ii}| > 1 \rightarrow$ Unilateral Potentially Stable Case
When $|S_{ii}| > 1$, it is possible that constant gain value G_i can take infinite value for the critical value of Γ_i, which is defined as

$$\Gamma_{i,c} = \frac{1}{S_{ii}} \tag{4.147}$$

g_i in Equation 4.145 can now take negative values due to the condition created by $|S_{ii}| > 1$. Hence, Γ_i should be chosen so that real part of the termination impedance is greater than the negative resistance at the point defined by $1/S_{ii}^*$. The process of drawing the gain circles is similar to the process described in the previous section.

4.2.4.5 Unilateral Figure of Merit

In practical cases, $S_{12} \neq 0$; so the criteria of unilateral design might not apply. However, it is possible to still use the unilateral design procedure by assuming that $S_{12} = 0$ with error. The error introduced by the approximation can be determined using the unilateral Figure of Merit. Consider the ratio of the transducer gain, G_T, to unilateral transducer gain, G_{TU}. This ratio can be expressed by

$$\frac{G_T}{G_{TU}} = \frac{1}{|1 - X|^2} \tag{4.148}$$

where

$$X = \frac{S_{12}S_{21}\Gamma_s\Gamma_L}{(1 - S_{11}\Gamma_s)(1 - S_{22}\Gamma_L)} \tag{4.149}$$

The transducer ratio given in Equation 4.148 is bounded by

$$\frac{1}{(1 + |X|)^2} < \frac{G_T}{G_{TU}} < \frac{1}{(1 - |X|)^2} \tag{4.150}$$

The maximum transducer gain occurs by simultaneous conjugate matching $\Gamma_s = S_{11}^*$ and $\Gamma_L = S_{22}^*$. So, the maximum error is introduced by

$$\frac{1}{(1 + U)^2} < \frac{G_T}{G_{T\max}} < \frac{1}{(1 - U)^2} \tag{4.151}$$

where

$$U = \frac{|S_{12}||S_{21}||S_{11}||S_{22}|}{(1 - |S_{11}^*|)(1 - |S_{22}^*|)} \tag{4.152}$$

In Equation 4.152, U is known as figure of merit.

Design Example

Design, simulate, and build a low-noise amplifier operating at 915 MHz using the small-signal amplifier design method. The amplifier uses ATF-54143 HEMT as transistor, and the S parameters of the device are given in Table 4.3. The amplifier is targeted to have a gain larger than 10 dB and a noise figure that is smaller than 2.5 dB with an operational bandwidth of 15%. ATF-5413 scattering parameters provided by the manufacturer are given in Table 4.3. The complete S parameter data is available on the manufacturer website [4].

Solution: We follow the small-signal amplifier design process flowchart illustrated in Figure 4.32. S parameters are provided by the manufacturer, Avago, for several frequencies. Hence, DC biasing circuit needs to be designed. This design problem will be simulated by advanced design system (ADS).

Step 1—Select Transistor
The transistor selected to design the low-noise amplifier is ATF-54143. This transistor was chosen as it has a wide frequency range, low noise figure, high gain, and low cost. ATF-54143 is a high-dynamic-range, low-noise, E-PHEMT-type transistor that comes with various packages such as SC-70 (SOT-343) surface mount plastic package. It is ideal for cellular/personal communication service (PCS) base stations, Multichannel Multipoint Distribution Service (MMDS), and other systems in the 450 MHz to 6 GHz frequency range. Before DC biasing is done, we need to obtain characteristics curve of the ATF-54143 transistor. The parameter values provided in ADS are compared with the transistor parameter values in the datasheet provided by the manufacturer and are in close agreement (Figure 4.44).

Step 2—Design DC Bias Circuit
The next stage was to design the DC biasing circuit for the transistor. S parameters were provided at specific DC biasing voltages and currents—four biasing pairs with independent S parameter values. From the datasheet, the S parameters, used to define the power flow, were defined over a wide range of frequencies. There is no specific S parameter set defined at 915 MHz, but S parameters at 900 MHz are provided earlier. Since

TABLE 4.3
Scattering Parameters for ATF-54143, $V_{DS} = 4$ V, $I_{DS} = 60$ mA

Freq.	S_{11}			S_{21}		S_{12}		S_{22}		MSG/ MAG
GHz	Mag.	Ang.	dB	Mag.	Ang.	Mag.	Ang.	Mag.	Ang.	vdB
0.1	0.99	−18.6	28.88	27.80	167.8	0.01	80.1	0.58	−12.6	34.44
0.5	0.81	−80.2	26.11	20.22	128.3	0.03	52.4	0.42	−52.3	28.29
0.9	0.71	−117.3	23.01	14.15	106.4	0.04	41.7	0.31	−73.3	25.49
1.0	0.69	−123.8	22.33	13.07	102.4	0.04	40.2	0.29	−76.9	25.14
1.5	0.64	−149.2	19.49	9.43	86.2	0.05	36.1	0.22	−89.4	22.76

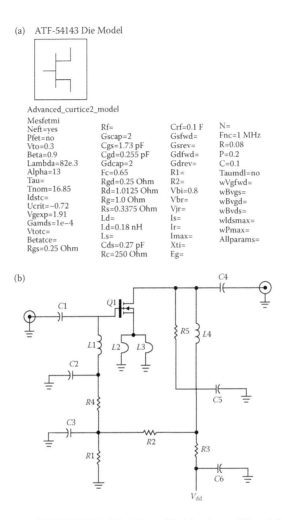

(a) ATF-54143 Die Model

Advanced_curtice2_model
Mesfetmi

Neft=yes	Rf=	Crf=0.1 F	N=
Pfet=no	Gscap=2	Gsfwd=	Fnc=1 MHz
Vto=0.3	Cgs=1.73 pF	Gsrev=	R=0.08
Beta=0.9	Cgd=0.255 pF	Gdfwd=	P=0.2
Lambda=82e.3	Gdcap=2	Gdrev=	C=0.1
Alpha=13	Fc=0.65	R1=	Taumdl=no
Tau=	Rgd=0.25 Ohm	R2=	wVgfwd=
Tnom=16.85	Rd=1.0125 Ohm	Vbi=0.8	wBvgs=
Idstc=	Rg=1.0 Ohm	Vbr=	wBvgd=
Ucrit=−0.72	Rs=0.3375 Ohm	Vjr=	wBvds=
Vgexp=1.91	Ld=	Is=	wldsmax=
Gamds=1e−4	Ld=0.18 nH	Ir=	wPmax=
Vtotc=	Ls=	Imax=	Allparams=
Betatce=	Cds=0.27 pF	Xti=	
Rgs=0.25 Ohm	Rc=250 Ohm	Eg=	

(b)

FIGURE 4.44   (a) ATF-54143 Die Model provided by Avago [4] and (b) DC biasing circuit for ATF-54143 [4].

the S parameter set provided is within 1.66% of the center frequency of 915 MHz, it is used in the design. The error observed was minimal with a maximum magnitude difference of 0.02 between the parameters provided at 1 GHz. A phase difference of less than 10° was noted, which also presented a minimal difference in real and imaginary values. With the minimal error in magnitude and phase difference presented earlier, the usage of the 900 MHz S parameters was satisfactory.

The biasing circuit that is recommended to be used for this transistor is shown in Figure 4.44b. The equations given to derive component values in Figure 4.44b are given in Table 4.4.

Before biasing is done, we need to obtain the characteristics curve of the ATF-54143 transistor. The transistor model parameters are given in Figure 4.45. The transistor

TABLE 4.4
DC Bias Circuit Design Equations

Component	Equation
V_{GS}—Voltage gate to source	
V_{DD}—Power supply voltage	
V_{DS}—Voltage drain to source; target parameter	
I_{DS}—Current drain to source; target parameter	
$I_{BB} = 10 * I_{gate\_leakage}$; $I_{gate\_leakage}$;	
$I_{gate\_leakage}$—Expected gate leakage current	
$C1, L1$	Input matching network
$C4, L4$	Output matching network
$C2, C5$	RF bypass
$C3, C6$	10 [nF]
$R1$	$R1 = \dfrac{V_{GS}}{I_{BB}}$
$R2$	$R2 = \dfrac{(V_{DS} - V_{GS}) * R1}{V_{GS}}$
$R3$	$R3 = \dfrac{V_{DD} - V_{DS}}{I_{DS} + I_{BB}}$
$R4$	Undefined by datasheet

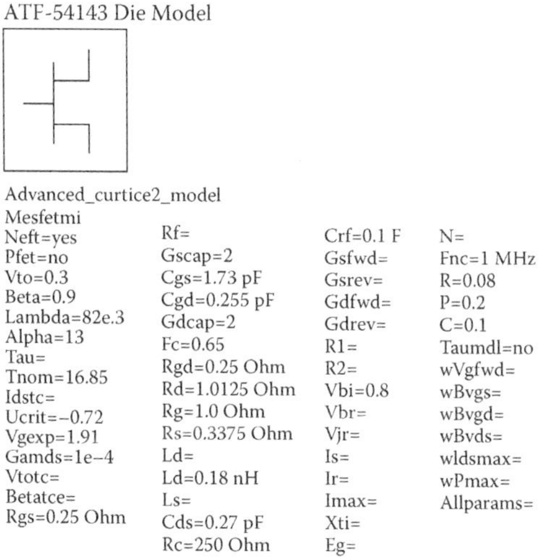

ATF-54143 Die Model

Advanced_curtice2_model

Mesfetmi			
Neft=yes	Rf=	Crf=0.1 F	N=
Pfet=no	Gscap=2	Gsfwd=	Fnc=1 MHz
Vto=0.3	Cgs=1.73 pF	Gsrev=	R=0.08
Beta=0.9	Cgd=0.255 pF	Gdfwd=	P=0.2
Lambda=82e.3	Gdcap=2	Gdrev=	C=0.1
Alpha=13	Fc=0.65	R1=	Taumdl=no
Tau=	Rgd=0.25 Ohm	R2=	wVgfwd=
Tnom=16.85	Rd=1.0125 Ohm	Vbi=0.8	wBvgs=
Idstc=	Rg=1.0 Ohm	Vbr=	wBvgd=
Ucrit=−0.72	Rs=0.3375 Ohm	Vjr=	wBvds=
Vgexp=1.91	Ld=	Is=	wldsmax=
Gamds=1e−4	Ld=0.18 nH	Ir=	wPmax=
Vtotc=	Ls=	Imax=	Allparams=
Betatce=	Cds=0.27 pF	Xti=	
Rgs=0.25 Ohm	Rc=250 Ohm	Eg=	

FIGURE 4.45 ATF-54143 Die Model provided by Avago.

I–V characteristics obtained in ADS are compared with the transistor *I–V* characteristic provided in the datasheet by the manufacturer (Figure 4.46). The *I–V* curves, which are obtained using ADS, are in close agreement as shown in Figure 4.47.

The biasing circuit simulation results matched with the *I–V* curves provided for the *S* parameter set in the datasheet when V_{DS} = 4 V and I_{DS} = 60 mA. The

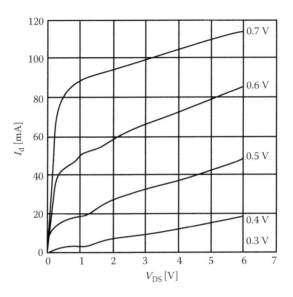

FIGURE 4.46 I_d vs V_{ds} provided by the manufacturer, Avago [4].

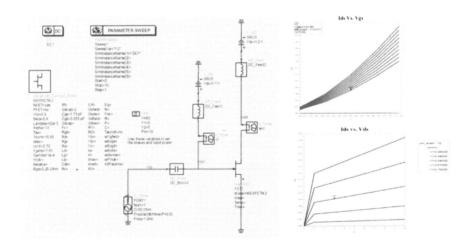

FIGURE 4.47   ATF-54143 I_d vs V_{GS} and I_d vs V_{DS} obtained from ADS.

component values of the final bias circuit and corresponding amplifier component values are illustrated in Table 4.5.

Get or Measure S Parameters within Frequency of Range
S parameters are given by the manufacturer as explained in Step 2 and are again given below as reference to be used for calculation.

$S_{11} = 0.71 < -117.3°, S_{12} = 0.04 < 41.7°, S_{21} = 14.15 < 106.4°$, and $S_{22} = 0.31 < -73.8°$.

TABLE 4.5

Final Component Values for the Amplifier

Part	Value	Qty
C1, C4	5.6 pF (0805)	2
C2, C5	18 pF(0805)	2
C3, C6	10 nF (0805)	2
L1	6.8 nF (LL2012)	1
L2, L3	Shorting strip	2
L4	8.2 nH (LL2012)	1
R1	4.7 kΩ (0805)	1
R2	33 kΩ (0805)	1
R3	27 Ω (0805)	1
R4	56 Ω (0805)	1
R5	330 Ω (0805)	1
Q1	ATF-54143	1

Step 3—Stability Analysis—$K > 1$ and $|\Delta| < 1$?
Based on the calculation using Equations 4.111 and 4.112,

$$k = \frac{1+|\Delta|^2 - |S_{11}|^2 - |S_{22}|^2}{2|S_{21}||S_{12}|} = 0.4739 < 1 \quad \text{and} \quad |\Delta| = |S_{22}S_{11} - S_{12}S_{21}| = 0.3697 < 1$$

Since, $k < 1$, the amplifier is potentially unstable. Since the transistor is potentially unstable, go to step 5.

Step 5—Draw Stability Circles
Because the amplifier is potentially unstable, we need to draw the stability circles. We need to identify whether the amplifier design can be done using unilateral design method. It is given that $S_{12} \neq 0$; so we need to determine the unilateral figure of merit and the maximum and minimum error associated with the assumption of unilateral design. From Equations 4.151 and 4.152,

$$\frac{1}{(1+U)^2} < \frac{G_T}{G_{Tmax}} < \frac{1}{(1-U)^2} \rightarrow -2.13\,[\text{dB}] < \frac{G_T}{G_{Tmax}} < 2.82\,[\text{dB}]$$

where

$$U = \frac{|S_{12}||S_{21}||S_{11}||S_{22}|}{\left(1-|S_{11}^*|\right)\left(1-|S_{22}^*|\right)}$$

So, the error will be bounded between −2.13 and 2.82 dB if we assume and proceed with unilateral design. For this design example, let us proceed with this error. However, it is important to note that this type of error is significant for applications where gain is critical.

The input and output stability circles are plotted using Equations 4.114 to 4.119. The radius and center point calculations are done with MATLAB and plotted as shown in Figure 4.48.

Step 5—Design Matching Network

We need to match the source impedance $Z_s = 50\,\Omega$, $Z_L = 50\,\Omega$ with $S_{11} = 0.71$ $< -117.3°$, and $S_{22} = 0.31 < -73.8°$. The unilateral design requires matching networks to be designed to follow $\Gamma_i = S_{ii}^*$. This is used to determine matching network components. Hence,

$$\Gamma_s = S_{11}^* = 0.71\angle 117.3° \rightarrow Z_{sm} = Z_o \frac{1+\Gamma_s}{1-\Gamma_s} = 11.5038 + 29.2718i$$

Similarly at the output,

$$\Gamma_L = S_{22}^* = 0.311\angle 73.8° \rightarrow Z_{Lm} = Z_o \frac{1+\Gamma_L}{1-\Gamma_L} = 48.9587 + 32.2482i$$

At the input, this requires to have series C and shunt L as an input matching network as shown in the Smith chart in Figure 4.49. After some tuning, when interfaced in the final circuit simulated by ADS, the component values are obtained as $C = 5.6$ [pF] and $L = 6.8$ [nH].

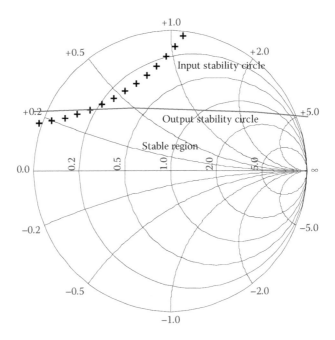

FIGURE 4.48   Input and output stability circles at 915 MHz.

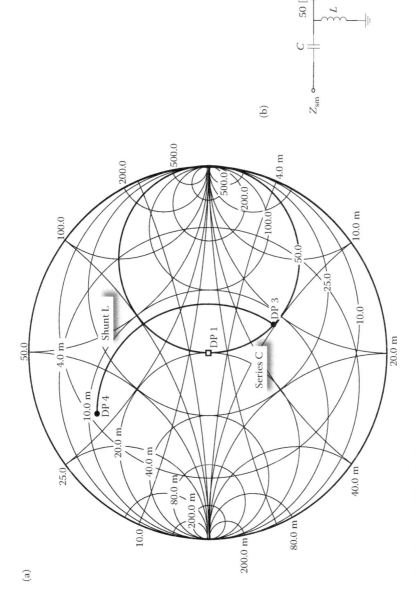

FIGURE 4.49 Input matching circuit.

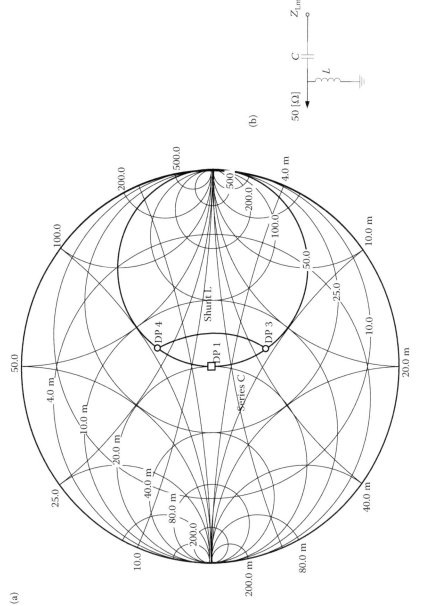

FIGURE 4.50 Output matching circuit.

Similarly, output matching network has series C and shunt as shown in the Smith chart in Figure 4.50. After some tuning, when interfaced in the final circuit simulated by ADS in Figure 4.48, the component values are obtained as $C = 5.6$ [pF] and $L = 8.2$ [nH].

Gain, Bandwidth, and Voltage Standing Wave Ratio

Gain, bandwidth, and voltage standing wave ratio of the amplifier are analyzed. The total theoretical transducer gain of the amplifier is found from Equation 4.127 as

$$G_T[dB] = G_s[dB] + G_o[dB] + G_L[dB] = 26.5 \ [dB]$$

The constant gain circles for source and load are given in Figures 4.51 and 4.52, respectively.

Tuning and Optimization

The final circuit in Figure 4.53 is simulated with ADS and optimized.

The gain obtained versus frequency illustrating also the bandwidth is given in Figure 4.54.

Prototyping

The simulated circuit shown in Figure 4.54 is built and tested. The prototype of the amplifier is shown in Figure 4.55.

The measured results are obtained using Agilent 4418B power meter with 10 dB coupler, which are illustrated in Table 4.6. Gain above 10 dB is obtained when the applied signal is greater than about 6 dBm.

The following is the MATLAB program used to do calculations and for plotting the gain and stability circles:

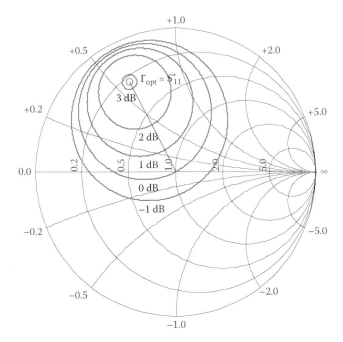

FIGURE 4.51 Constant gain circles at the input.

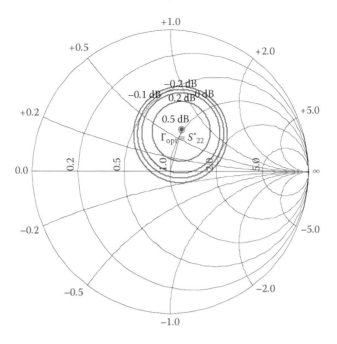

FIGURE 4.52 Constant gain circles at the output.

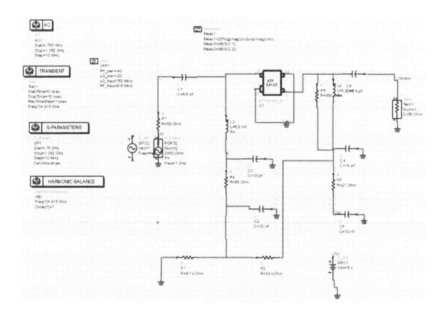

FIGURE 4.53 Simulation of the final circuit with ADS.

(a)

Frequency	AC1. meas 1
750.0 MHz	16. 937
760.0 MHz	17. 317
770.0 MHz	17. 617
780.0 MHz	17. 842
790.0 MHz	17. 998
800.0 MHz	18. 091
810.0 MHz	18. 131
820.0 MHz	18. 126
830.0 MHz	18. 085
840.0 MHz	18. 015
850.0 MHz	17. 923
860.0 MHz	17. 815
870.0 MHz	17. 694
880.0 MHz	17. 565
890.0 MHz	17. 430
900.0 MHz	17. 292
910.0 MHz	17. 152
920.0 MHz	17. 011
930.0 MHz	16. 871
940.0 MHz	16. 732
950.0 MHz	16. 595
960.0 MHz	16. 460
970.0 MHz	16. 328
980.0 MHz	16. 198
990.0 MHz	16. 070
1.000 GHz	15. 945
1.010 GHz	15. 823
1.020 GHz	15. 704
1.030 GHz	15. 586
1.040 GHz	15. 472

(b)

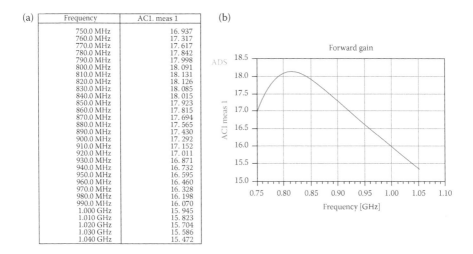

FIGURE 4.54 ADS simulation results of the final circuit with ADS.

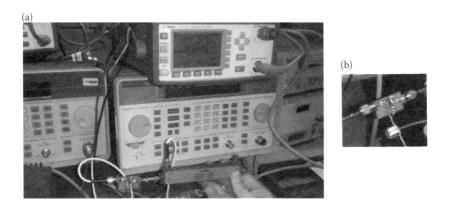

FIGURE 4.55 The prototype of the low-noise amplifier built and tested.

TABLE 4.6
Power Meter Gains

Frequency [MHz]	Power In [dBm]	Power Out [dBm]
915	4	12.1
915	5	13.98
915	6	15.6
915	8	17.5
915	10	17.8

MATLAB Script to Design Small-Signal Amplifier

```
close all;    % close all opened graphs
clear all;

smith_chart;

prompt = {'Zo','S11Mag','S11Ang','S12Mag','S12Ang','S21Mag','S
21Ang','S22Mag','S22Ang' };
dlg_title = 'Scattering Parameters ';
num_lines = 1;
def =
{'50','0.71','-117.3','0.04','41.7','14.15','106.4','0.31','-
73.8'};
answer = inputdlg(prompt,dlg_title,num_lines,def, 'on');

%convert the strings received from the GUI to numbers
valuearray=str2double(answer);

%Give variable names to the received numbers
Zo=valuearray(1);
S11Mag=valuearray(2);
S11Ang=valuearray(3);
S12Mag=valuearray(4);
S12Ang=valuearray(5);
S21Mag=valuearray(6);
S21Ang=valuearray(7);
S22Mag=valuearray(8);
S22Ang=valuearray(9);




%    Enter angle in degrees, this converts to complex.

S11rad = S11Ang*pi/180;%radian conversion for complex notation
S12rad = S12Ang*pi/180;
S21rad = S21Ang*pi/180;
S22rad = S22Ang*pi/180;
S = [S11Mag*exp(1i*S11rad),S12Mag*exp(1i*S12rad); S21Mag*exp(1
i*S21rad),S22Mag*exp(1i*S22rad)];


input_stability(S, '+k');
output_stability(S, '-r');


delta=S(1,1)*S(2,2)-S(1,2)*S(2,1);
fprintf('Delta');
disp(abs(delta));
```

```
K=(1+(abs(delta))^2-(abs(S(1,1)))^2-(abs(S(2,2))^2))/
(2*abs(S(1,2)*S(2,1)));
fprintf('\n Roulette K factor:');
disp(K);
if K > 1 && delta < 1
    fprintf('Unconditionally Stable');
else
    fprintf('Ptentially Unstable');
end


%Input Stabilty Circle Parameters

Cs=conj(S(1,1)-delta*conj(S(2,2)))/
(abs(S(1,1))^2-abs(delta)^2);
Rs=abs(S(1,2)*S(2,1)/(abs(S(1,1))^2-abs(delta)^2));

%%Output Stabilty Circle Parameters

Cl=conj(S(2,2)-delta*conj(S(1,1)))/
(abs(S(2,2))^2-abs(delta)^2);
Rl=abs(S(1,2)*S(2,1)/(abs(S(2,2))^2-abs(delta)^2));

%Optimum Termination Impedances

GammaS=conj(S(1,1));
GammaL=conj(S(2,2));

Zsm=Zo*((1+GammaS)/(1-GammaS)); %Source Impedance
Zlm=Zo*((1+GammaL)/(1-GammaL)); %Load Impedance

%Unilateral Figure of Merit, max and min errors

U=(S11Mag*S12Mag*S21Mag*S22Mag)/((1-S11Mag^2)*(1-S22Mag^2));
X=(S(1,2)*S(2,1)*GammaS*GammaL)/
((1-S(1,1)*GammaS*(1-S(1,1)*GammaL)));
Max_Error=1/((1-U)^2);
Max_Error_dB=10*log10(Max_Error);
Min_Error=1/((1+U)^2);
Min_Error_dB=10*log10(Min_Error);
FoM=1/(abs(1-X))^2; % Ratio Gt/Gtu

% Calculation of GammaIn and GammaOut

GammaOut=S(2,2)+((S(1,2)*S(2,1)*GammaS)/(1-S(1,1)*GammaS));
GammaIn=S(1,1)+((S(1,2)*S(2,1)*GammaL)/(1-S(2,2)*GammaL));



% Source Side Gain Calculation
```

```
Gs_max=1/(1-abs(S(1,1))^2);
Gs_max_dB=10*log10(Gs_max);
Gs=(1-abs(GammaS)^2)/((abs(1-S(1,1)*GammaS))^2);
gs=Gs/Gs_max;



% Load Side Gain Calculation

Gl_max=1/(1-(abs(S(2,2)))^2);
Gl_max_dB=10*log10(Gl_max);
Gl=(1-(abs(GammaL))^2)/((abs(1-S(2,2)*GammaL))^2);
gl=Gl/Gl_max;

% Transistor Gain

G0=abs(S(2,1))^2;
G0_dB=10*log10(G0);

%Total Gain in dB

Gtu_dB=Gs_max_dB+Gl_max_dB+G0_dB;


%Source gain circles

smith_chart;

%Straight line connecting GammaS and the origin
hold on;
plot([0 real(GammaS)],[0 imag(GammaS)],'b');
plot(real(GammaS),imag(GammaS),'bo');

% specify the angle for the constant gain circles
a=(0:360)/180*pi;

gs_db=[-1 0 1 2 3]; % range of desired gains
gs=exp(gs_db/10*log(10))/Gs_max; % convert from dB to normal
units

for n=1:length(gs)
    dg=gs(n)*conj(S(1,1))/(1-abs(S(1,1))^2*(1-gs(n)));
    rg=sqrt(1-gs(n))*(1-abs(S(1,1))^2)/
(1-abs(S(1,1))^2*(1-gs(n)));
    plot(real(dg)+rg*cos(a),imag(dg)+rg*sin(a),'r','linewi
dth',2);
    text(real(dg)-0.05,imag(dg)+rg+0.05,strcat('\
bf',sprintf('%gdB',gs_db(n))));

end;
```

```
text(real(GammaS)-0.05,imag(GammaS)-0.06,'\bf\
Gamma_{opt}=S_{11}^*');



% Load Gain Circles

smith_chart;


%draw a straight line connecting Gs_opt and the origin
hold on;
plot([0 real(GammaL)],[0 imag(GammaL)],'b');
plot(real(GammaL),imag(GammaL),'bo');

% specify the angle for the constant gain circles
a=(0:360)/180*pi;

%plot source gain circles
gl_db=[-0.2 -0.1 0 0.2 0.5]; % range of desired gains
gl=exp(gl_db/10*log(10))/Gl_max; % convert from dB to normal
units

for n=1:length(gl)
    dl=gl(n)*conj(S(2,2))/(1-abs(S(2,2))^2*(1-gl(n)));
    rl=sqrt(1-gl(n))*(1-abs(S(2,2))^2)/
(1-abs(S(2,2))^2*(1-gl(n)));
    plot(real(dl)+rl*cos(a),imag(dl)+rl*sin(a),'r','linewi
dth',2);
    text(real(dl)-0.05,imag(dl)+rl+0.05,strcat('\
bf',sprintf('%gdB',gl_db(n))));
end;

text(real(GammaL)-0.05,imag(GammaL)-0.06,'\bf\
Gamma_{opt}=S_{22}^*');


fprintf('\n Load Stability circle radius:');
disp(Rl);
fprintf('\n Load Stability circle center:');
disp(Cl);
fprintf('\n Source Stability circle radius:');
disp(Rs);
fprintf('\n Source Stability circle center:');
disp(Cs);
fprintf('\n GammaS:');
disp(GammaS);
fprintf('\n GammaL:');
disp(GammaL);
fprintf('Unilateral Figure of Merit:');
disp(U);
```

```
fprintf('\n Max Error: ');
disp(Max_Error_dB);
fprintf('\n Min Error: ');
disp(Min_Error_dB);
fprintf('\n GT/GTU:');
disp(FoM);
fprintf('\n Zs:');
disp(Zsm);
fprintf('\n ZL:');
disp(Zlm);
fprintf('\n Gtu in dB:');
disp(Gtu_dB);
fprintf('\n GammaIn:');
disp(GammaIn);
fprintf('\n GammaOut:');
disp(GammaOut);
```

REFERENCES

1. G. Gozales, *Microwave Transistor Amplifiers: Analysis and Design*, 2nd ed., Pearson, 1996.
2. G. D. Vendelin, A. M. Pavio, and U. L. Rohde, *Microwave Circuit Design Using Linear and Nonlinear Techniques*, 2nd ed., Wiley, 2005.
3. HP test and application note 95-1, S-Parameter techniques, February 1967.
4. Transistor datasheet. https://www.modelithics.com/models/Vendor/Avago/ATF-54143.pdf.

5 Linear Amplifier Design and Implementation

5.1 LARGE-SIGNAL RF AMPLIFIER DESIGN TECHNIQUES

Small-signal analysis methods are applicable when the amount of signal applied is low enough not to change the linearity of the device. This is part of the reason why manufacturers of active devices, e.g., transistors, provide designers with small-signal parameters, such as S parameters, for specific cases of low-signal applications. However, many practical applications require signal levels that are high, which changes the characteristics of the device. Hence, small-signal parameters become no longer applicable. There are large-signal parameters that can be used to design large-signal amplifiers. However, large-signal S parameters are not easy to measure and hence not being used in practice.

In practice, the accepted method to design high-power radio frequency (RF) amplifier is using the large-signal input and output impedances provided by the manufacturer for RF transistors. Load-pull measurement method is then used to characterize the gain, power, and efficiency characteristics of the devices at the operating frequency within bandwidth [1–7].

Load-pull measurement method can be understood better by studying the basic Class A amplifier circuit given in Figure 5.1. The calculation of the optimum load resistance, R_{opt}, can be illustrated from the DC load line for Class A amplifier shown in Figure 5.2.

The calculation of the optimum load resistance, R_{opt}, can be found from the DC load line theory as

$$R_{opt} = Z_L = \frac{V_{dc} - V_{knee}}{I_{max}/2} = \frac{V_{dc} - V_{knee}}{I_{dc}} \tag{5.1}$$

The optimum output power and drain efficiency can then be calculated as

$$P_{opt} = \frac{(V_{dc} - V_{knee})I_{dc}}{2} \tag{5.2}$$

and

$$\eta_{max} = \frac{P_{opt}}{P_{dc}} = \frac{(V_{dc} - V_{knee})}{2V_{dc}} \tag{5.3}$$

For an ideal case, $V_{knee}=0$, and hence the maximum efficiency is 50% as expected for Class A amplifiers. When $V_{knee}=0$, Equation 5.2 can be written as

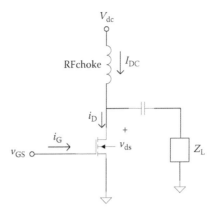

FIGURE 5.1   Load-pull measurement circuit.

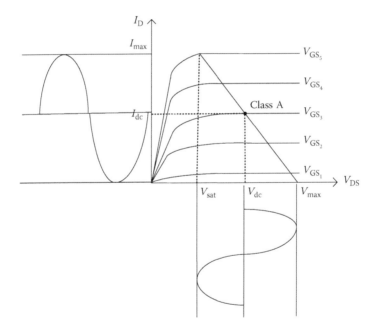

FIGURE 5.2   DC load line for Class A amplifier for R_{opt} for calculation.

$$P_{opt} = \frac{V_{dc} I_{dc}}{2} = \frac{V_{dc}^2}{2 R_{opt}} = \frac{1}{2} I_{dc}^2 R_{opt} \tag{5.4}$$

R_{opt} on the Smith chart can be illustrated as shown in Figure 5.3.

When there is a mismatch, it will have an effect on the output power and efficiency of the power amplifier (PA). This effect can be found by using contours on the Smith chart. Let us define a quantity p, which is a power reduction factor. The power

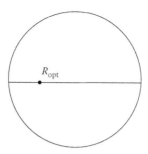

FIGURE 5.3   Illustration of R_{opt} on the Smith chart.

output can then be represented by P_{opt}/p. The possible resistive solutions that can be written for this mismatch are as follows:

Resistive Solution 1

$$pR_{\text{opt}} = \begin{cases} I_m = \dfrac{V_{\text{dc}}}{pR_{\text{opt}}} \\[3mm] V_m = V_{\text{dc}} \end{cases} \rightarrow P_o = \frac{1}{2}\frac{V_{\text{dc}}^2}{pR_{\text{opt}}} = \frac{P_{\text{opt}}}{p} \tag{5.5}$$

Resistive Solution 2

$$\frac{R_{\text{opt}}}{p} = \begin{cases} I_m = \dfrac{I_{\max}}{2} = I_{\text{dc}} \\[3mm] V_m = \dfrac{I_{\text{dc}}R_{\text{opt}}}{p} \end{cases} \rightarrow P_o = \frac{1}{2}\frac{I_{\text{dc}}^2 R_{\text{opt}}}{p} = \frac{P_{\text{opt}}}{p} \tag{5.6}$$

Based on the results, the possible impedance values are

$$R_{\text{L}} = \frac{R_{\text{opt}}}{p} \tag{5.7a}$$

$$R_{\text{H}} = pR_{\text{opt}} \tag{5.7b}$$

These two possible solutions can be illustrated using the Smith chart given in Figure 5.4.

We can now begin analysis of two different cases for the upper and lower limits: $R=R_{\text{L}}$ and $R=R_{\text{H}}$.

$$\text{Case 1}:\ R_{\text{L}} = \frac{R_{\text{opt}}}{p}$$

When $R=R_{\text{L}}$, we need to add series reactance jX to increase the voltage swing without affecting current, because power is limited by the current for this case. This can be illustrated in Figure 5.5.

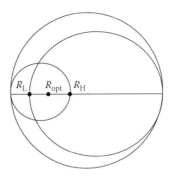

FIGURE 5.4   Illustration of low and high impedance limits on the Smith chart.

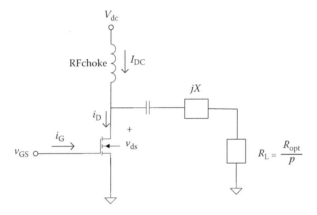

FIGURE 5.5   Adding series reactance when there is low impedance, R_L.

The boundary for the added reactance is identified as $\pm X_\mathrm{m}$, and it can be found from the following equation:

$$\left| R_\mathrm{L} + jX_\mathrm{m} \right| = R_\mathrm{opt} \tag{5.8}$$

Case 1: $R_\mathrm{H} = pR_\mathrm{opt}$

When $R = R_\mathrm{L}$, we need to add shunt component with susceptance, jB, to increase the current swing without affecting the voltage, because power is limited by the voltage for this case. This can be illustrated in Figure 5.6.

The boundary that current can be increased will be defined by the following equation:

$$\left| G_\mathrm{L} + jB_\mathrm{m} \right| = \left| \frac{1}{R_\mathrm{opt}} \right| \tag{5.9}$$

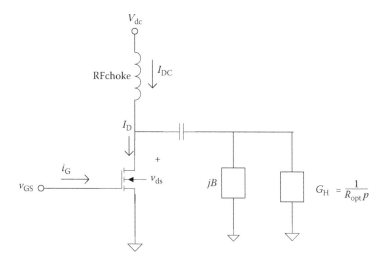

FIGURE 5.6   Adding shunt susceptance when there is high impedance, G_H.

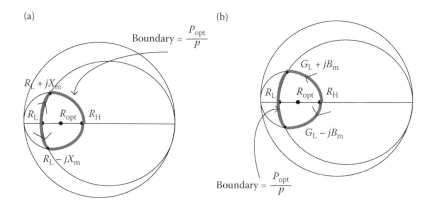

FIGURE 5.7   Constant power contour using load pull technique; (a) Case 1 and (b) Case 2.

Therefore, load-pull contours are identified by following constant r or g circles on the Smith chart as shown in Figure 5.7. The intersection of constant resistance and conductance are the limits that are identified by Equations 5.8 and 5.9, X_m and B_m, respectively. As a result, the contour obtained shows the constant power for the specified case.

5.2   PUSH–PULL AMPLIFIER CONFIGURATION

The output power capacity of an RF amplifier can be increased in several ways. One of the ways is to use higher voltage and current-rated active devices for the frequency of operation. However, this has many implications such as availability of devices, increased size, and other implementation issues. When use of higher voltage and current-rated devices is not feasible, the output power capacity can also be increased by using push–pull configuration and/or parallel transistor configuration.

Practical RF PA design uses push–pull configuration widely to meet the demand of high output power. The input drive signal for push–pull amplifiers is out phased by 180° using transformers such as input balun. Output balun is used at the output to combine the out-phased amplifier output signal and double the RF power. RF input voltage signal at operational frequency ignoring the phases and assuming ideal conditions for simplicity can be expressed as

$$v_s(t) = 4V_m \sin(\omega t) \tag{5.10}$$

Assuming the matched impedance case, the signal at the input of the balun is then

$$v_{in}(t) = \frac{v_s(t)}{2} = 2V_m \sin(\omega t) \tag{5.11}$$

The signal at the output of the balun will be equally split and phased by 180°. These signals at the input of the amplifiers can be represented as

$$v_1(t) = V_m \sin(\omega t) \tag{5.12}$$

$$v_2(t) = V_m \sin(\omega t + \theta) = -V_m \sin(\omega t) \tag{5.13}$$

where $\theta = 180°$.

The signals $v_1(t)$ and $v_2(t)$ will be amplified by the amplifiers by their corresponding gains, A_1 and A_2, respectively. The signals at the output of the amplifiers are then equal to

$$v_1(t) = A_1 V_m \sin(\omega t) \tag{5.14}$$

$$v_2(t) = -A_2 V_m \sin(\omega t) \tag{5.15}$$

The amplifier output signals are then combined via output balun. The final load signal is

$$v_L(t) = v_1(t) - v_2(t) = (A_1 + A_2)V_m \sin(\omega t) \tag{5.16}$$

The illustration of the push–pull amplifier with waveforms is given in Figure 5.8.

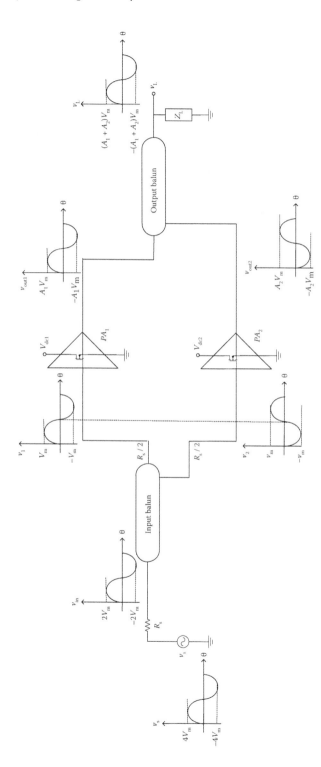

FIGURE 5.8 Implementation of push–pull amplifiers.

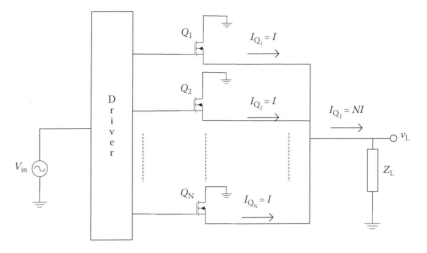

FIGURE 5.9 Parallel connection of transistors for RF amplifiers.

5.3 PARALLEL TRANSISTOR CONFIGURATION

RF power output in the amplifier system can be increased by also paralleling the transistors. When transistors are paralleled, the current is increased proportionally to the number of the transistors used under ideal conditions. This can be illustrated in Figure 5.9.

5.4 PA MODULE COMBINERS

RF output power can be increased further up to several kWs by combining individual identical PA modules. These PA modules are usually combined via Wilkinson-type power combiner. The typical PA module combining technique is shown via a two-way power combiner in Figure 5.10.

5.5 LINEAR AMPLIFIERS

In this section, amplifier classes such as Class A, Class B, and Class AB for linear mode of operation will be discussed.

When the transistor is operated as a dependent current source, the conduction angle, 2θ, is used to determine the class of the amplifier shown in Figure 5.11. The conduction angle varies up to 2π based on the amplifier class. The use of transistor as a dependent current source representing a linear mode of operation was briefly discussed in Chapter 1 and is shown in Figure 5.11 as reference again.

The conduction angles, bias, and quiescent points for linear amplifier are shown in Figure 5.12 and are illustrated in Table 5.1 as reference. As it was discussed in Chapter 1, conduction angle, θ, is defined as the duration of the period in which the given transistor is conducting, and the full cycle of conduction was considered to be 360°.

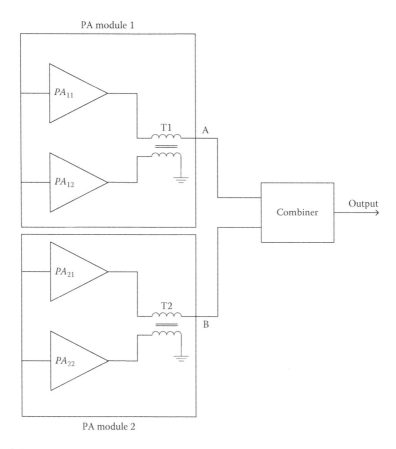

FIGURE 5.10   Power combiner for PA modules.

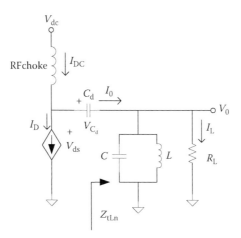

FIGURE 5.11   Equivalent circuit representation of linear amplifier mode of operation.

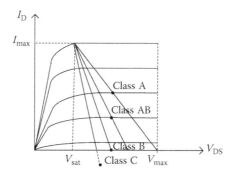

FIGURE 5.12   Load lines and bias points for linear amplifiers.

TABLE 5.1

Conduction Angles, Bias, and Quiescent Points for Linear Amplifiers

Class	Bias Point	Quiescent Point	Conduction Angle
A	0.5	0.5	2π
AB	0–0.5	0–0.5	π–2π
B	0	0	π
C	<0	0	0–π

In our analysis, metal-oxide field-effect transistor (MOSFET) is used as an active device in the amplifier circuit. The ideal MOSFET transistor transfer characteristics are illustrated in Figure 5.13. There is maximum drain current level for a corresponding gate-source voltage (V_{gs}) that an MOSFET conducts. In the cutoff region, the gate-source voltage, V_{gs}, is less than threshold voltage or saturation voltage, V_{sat}, and the device is an open circuit or off. In the Ohmic region, the device acts as a resistor with an almost constant on-resistance R_{dsON} and is equal to the ratio of drain voltage, V_{DS}, and the drain current, I_D. In the linear mode of operation, the device operates in the active region, where I_D is a function of the gate-source voltage V_{gs} and defined by

$$I_D = K_n \left(V_{gs} - V_{th}\right)^2 = g_m \left(V_{gs} - V_{th}\right) \qquad (5.17)$$

where K_n is a parameter depending on the temperature and device geometry, and g_m is the current gain or transconductance. When V_{DS} is increased, the positive drain potential opposes the gate voltage bias and reduces the surface potential in the channel. The channel inversion-layer charge decreases with increasing V_{DS} and, ultimately, becomes zero when the drain voltage equals to $V_{gs} - V_{th}$. This point is called the "channel pinch-off point," where the drain current becomes saturated.

When operational drain current (I_D) at a given V_{gs} goes above the Ohmic or linear region "knee or saturation point", any further increase in drain current results in a significant rise in drain-source voltage (V_{DS}) as shown in Figure 5.13. This results

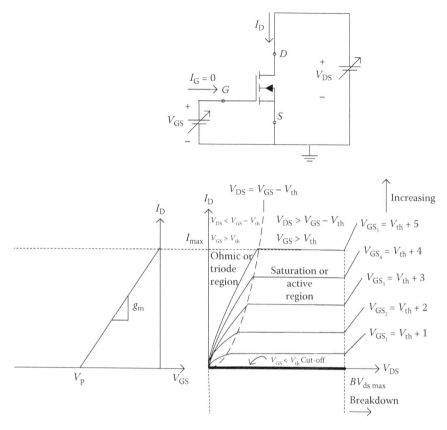

FIGURE 5.13 Transfer characteristics of ideal FET device, N channel Metal Oxide Semiconductor (NMOS).

in a rise in conduction loss. If power dissipation is high and over the limit that the transistor can handle, then the device may catastrophically fail. When V_{gs} is less than the threshold or pinch-off voltage, V_P, then the device does not conduct, and I_D becomes 0. As V_{gs} increases, the transistor enters the saturation or active region, and I_D increases in a nonlinear fashion. It will remain almost a constant until the transistor gets into breakdown region. The characteristics of I_D versus V_{gs} in saturation region are illustrated in Figure 5.14. The simplified device model for MOSFET in each region is illustrated in Figure 5.15.

Example 5.1

State whether the following transistor in Figure 5.16 is operating in saturation region. What should be the value of gate voltage, V_G, for the transistor to operate in saturation region? Assume $V_t = 1$ V.

Solution: The condition to operate in saturation region is

$$V_{gs} > V_t$$

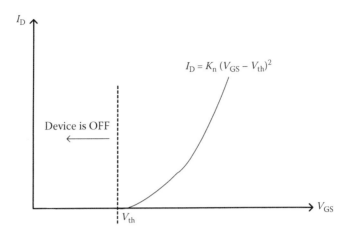

FIGURE 5.14 I_D versus V_{gs} in saturation region for ideal NMOS device.

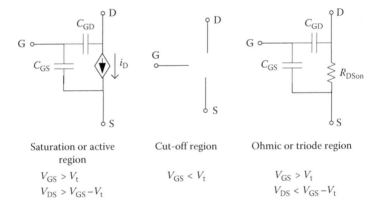

FIGURE 5.15   Simplified large-signal model for NMOS FET device.

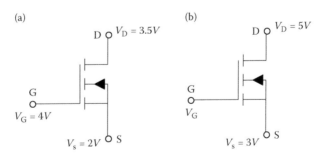

FIGURE 5.16   MOSFET illustration for Example 5.1.

$$V_{DS} > V_{gs} - V_t$$

1. $V_{gs} = 4 - 2 = 2\,[V] > V_t = 1\,[V]$ and $V_{DS} = 3.5 - 2 = 1.5\,[V] > V_{gs} - V_t = 2 - 1 = 1\,[V]$

   Therefore, the first transistor is operating in saturation region.
2. From the first condition, for transistor to operate in saturation region, $V_{gs} > 1\,[V]$. That requires gate voltage $V_G > 4\,[V]$. In addition, it is required that $V_{gs} < V_{DS} + V_t = 2 + 1 = 3\,[V]$. The overall solution is

$$1 < V_{gs} < 3$$

Since source voltage is $V_S = 2\,V$,

$$1 < V_G - 2 < 3 \text{ or } 3 < V_G < 5$$

5.5.1 CONVENTIONAL AMPLIFIERS: CLASSES A, B, AND C

For amplifier Classes A, B, and C, the transistor can be modeled as the voltage-dependent current source (Figure 5.17). Most of the time, the class of these amplifiers are also known as conventional amplifiers.

The current i_D flowing through the device, drain-to-source voltage, and the voltage applied at the gate to source of the transistor in Figure 5.17 are

$$i_D = I_{dc} + I_m \cos(\theta) \tag{5.18}$$

$$v_{DS} = V_{dc} - V_m \cos(\theta) \tag{5.19}$$

$$v_{gs} = V_t + V_{gsm} \cos(\theta) \tag{5.20}$$

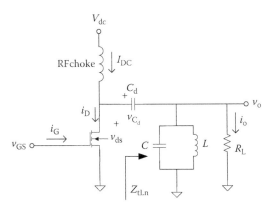

FIGURE 5.17   Conventional RF power amplifier classes: Class A, B, and C.

where $\theta = \omega t$. The DC components of the drain current and drain-to-source voltage are equal to I_{dc} and V_{dc}, and AC component is $I_m \cos(\theta)$, $-V_m \cos(\theta)$ and given as

$$I_D = I_{dc}, \quad V_{DS} = V_{dc} \tag{5.21}$$

$$i_D = I_m \cos(\theta), \quad v_{DS} = -V_m \cos(\theta) \tag{5.22}$$

Fourier integrals can be used to determine the DC power and output of the amplifier. Drain current, i_D, can be represented using Fourier series expansion with sine or cosine functions, which are harmonically related as

$$i_D(t) = I_o + \sum_{n=1}^{\infty} I_{an} \cos n\omega_o t + I_{bn} \sin n\omega_o t \tag{5.23}$$

$$v_{DS}(t) = V_o + \sum_{n=1}^{\infty} V_{an} \cos n\omega_o t + V_{bn} \sin n\omega_o t \tag{5.24}$$

We can combine the odd and even harmonic coefficients into single cosine (or sine) to give

$$i_D(t) = I_o + \sum_{n=1}^{\infty} I_n \cos\left(n\omega_o t + \alpha_n\right) \tag{5.25}$$

$$v_{DS}(t) = V_o + \sum_{n=1}^{\infty} V_n \cos\left(n\omega_o t + \beta_n\right) \tag{5.26}$$

where

$$I_n = \sqrt{I_{an}^2 + I_{bn}^2} \tag{5.27a}$$

and

$$V_n = \sqrt{V_{an}^2 + V_{bn}^2} \tag{5.27b}$$

and

$$\alpha_n = \tan^{-1}\left(\frac{I_{bn}}{I_{an}}\right) \tag{5.28a}$$

and

$$\beta_n = \tan^{-1}\left(\frac{V_{bn}}{V_{an}}\right) \tag{5.28b}$$

The impedance at the transistor load line of the transistor is found from

$$Z_{tLn} = \frac{V_n e^{j\alpha}}{I_n e^{j\beta}} = |Z_{tLn}| e^{j\phi_n} \tag{5.29}$$

where $\phi_n = (\alpha_n - \beta_n)$. The Fourier coefficients I_o, I_{an}, and I_{bn} for drain current are calculated from

$$I_o = \frac{1}{T} \int_{-T/2}^{T/2} i_D(t)\,dt \tag{5.30a}$$

$$I_{an} = \frac{2}{T} \int_{-T/2}^{T/2} i_D(t) \cos(k\omega_o t)\,dt \tag{5.30b}$$

$$I_{bn} = \frac{2}{T} \int_{-T/2}^{T/2} i_D(t) \sin(k\omega_o t)\,dt \tag{5.30c}$$

where $\omega_o = 2\pi/T$. The fundamental component in Equation 5.30 is obtained when $n = 1$, and DC components are found from Equation 5.30a as follows:

$$I_o = \frac{1}{2\pi} \int_{-\pi}^{\pi} \left(I_{dc} + I_m \cos(\theta)\right) d\theta = I_{dc} \tag{5.31a}$$

$$I_{an} = \frac{1}{\pi} \int_{-\pi}^{\pi} \left(I_{dc} + I_m \cos(\theta)\right) \cos(n\theta)\, d\theta = I_m \quad \text{when } n = 1 \text{ otherwise } I_{an} = 0 \tag{5.31b}$$

$$I_{bn} = \frac{1}{\pi} \int_{-\pi}^{\pi} \left(I_{dc} + I_m \cos(\theta)\right) \sin(n\theta)\, d\theta = 0 \text{ for all } n \tag{5.31c}$$

The same analysis and derivation can also be repeated for drain-to-source voltage. Hence, the DC and the fundamental components of the drain current and drain-to-source voltage at the resonant frequency, f_o, are

$$I_o = I_{dc}, \quad V_o = V_{dc} \tag{5.32}$$

$$I_1 = I_m, \quad V_1 = V_m \tag{5.33}$$

DC power, P_{dc} from supply is then calculated from $P_{dc} = V_{dc} I_{dc}$ (5.34)

The power delivered from device to output is represented by P_o and is calculated at the fundamental frequency, which is the resonant frequency of the LC network obtained from Equation 5.33 as

$$P_o = \frac{1}{2} V_m I_m \tag{5.35}$$

The more general expression when operational frequency is not equal to resonant frequency for the output power can be found at the fundamental and harmonic frequencies as

$$P_{on} = \frac{1}{2} V_n I_n \cos(\phi_n), \quad n = 1, 2, \ldots \tag{5.36}$$

The transistor dissipation is calculated using

$$P_{diss} = \frac{1}{T} \int_0^T v_{DS}(t) i_D(t) \, dt \tag{5.37}$$

which is also equal to

$$P_{diss} = P_{dc} - \sum_{n=1}^{\infty} P_{o,n} \tag{5.38}$$

The drain efficiency is the ratio of the output power to DC supply power and is calculated using Equations 5.34 and 5.36 as

$$\eta = \frac{P_{o,n}}{P_{dc}} = \frac{P_{o,n}}{P_{diss} + P_{o,n}} \tag{5.39}$$

The maximum efficiency is obtained when

$$P_{diss} + \sum_{n=2}^{\infty} P_{o,n} = 0 \tag{5.40}$$

Then, the maximum efficiency from Equation 5.39 is found to be $\eta_{max} = 100\%$.

The drive input power is calculated using Equation 5.20 for gate-to-source voltage, v_{gs}, and gate current, i_G, from

$$P_G = \frac{1}{T} \int_0^T v_{gs}(t) i_G(t) \, dt \tag{5.41}$$

5.6 CLASS A AMPLIFIERS

For amplifier Classes A, B, and C, the transistor can be modeled as the voltage-dependent current source (Figure 5.18). Most of the time, the class of these amplifiers are also known as conventional amplifiers.

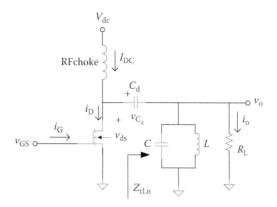

FIGURE 5.18 Equivalent circuit for RF power amplifier classes: Class A, B, and C.

The bias point in Class A mode of operation is selected at the center of I–V curve between the saturation voltage, V_{sat}, and maximum operational transistor voltage, V_{max}, whereas DC current for Class A amplifier is biased between 0 and the maximum allowable current, I_{max}. The conduction angle for the transistor for Class A operation is 2π, which means that the transistor conducts the entire RF cycle. Current and voltage waveforms for Class A amplifier are illustrated in Figure 5.19.

High operational current exists with high voltage because DC bias point is located in the middle of I–V characteristics. This results in high dissipation. The Class A amplifier does not need an LC resonator network shown in Figure 5.18, because the output of the amplifier is sinusoidal. Nonetheless, the tuning network is usually

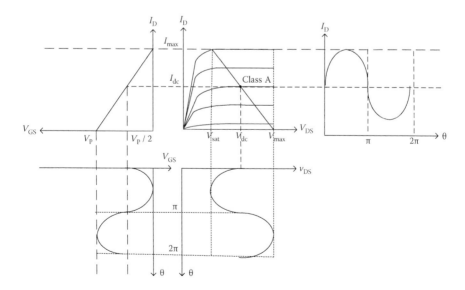

FIGURE 5.19 Class A amplifier waveforms.

implemented in practice due to nonideal device characteristics. Voltage and current for Class A mode of operation can be expressed as

$$i_D = I_{dc} + I_m \cos(\theta) \tag{5.42}$$

$$v_{DS} = V_{dc} - V_m \cos(\theta) \tag{5.43}$$

$$v_{gs} = V_t + V_{gsm} \cos(\theta) \tag{5.44}$$

The output current, i_o, and output voltage, v_o, are

$$i_o = I_m \cos(\theta) \tag{5.45}$$

$$v_o = -V_m \cos(\theta) = -I_m R_L \cos(\theta) \tag{5.46}$$

Therefore, Equations 5.42 and 5.43 can be set in the following form:

$$i_D = I_{dc} + i_o(\theta) \tag{5.47}$$

$$v_{DS} = V_{dc} + v_o(\theta) \tag{5.48}$$

When Equations 5.42 and 5.43 with Equations 5.45 through 5.46 are solved, we obtain the load line equation as

$$i_D = -\frac{v_{ds}}{R_L} + \left(I_{dc} + \frac{1}{R_L} V_{dc} \right) \tag{5.49}$$

Power and efficiency are given in Figure 5.20.

5.7   CLASS B

Transistor power dissipation due to its 360° conduction angle for Class A amplifiers significantly limits the amplifier RF output power capacity. The power dissipation in the active device can be reduced if the device is biased to conduct less than the full RF period. The transistor is turned on only one-half of the cycle, and as a result the conduction angle for Class B amplifiers is $\theta = 180°$. Class B amplifiers can be implemented as a single-ended amplifier when narrowband is required, or as transformed coupled push–pull configuration when high linear output power is desired. The typical load line waveforms are illustrated in Figure 5.21.

In practice, Class B amplifiers are implemented in push–pull configuration as shown in Figure 5.22.

In Figure 5.22, input and output transformers, T_1 and T_2, are ideal, and the output current, i_o, and output voltage, v_o, are sinusoidal and can be expressed as

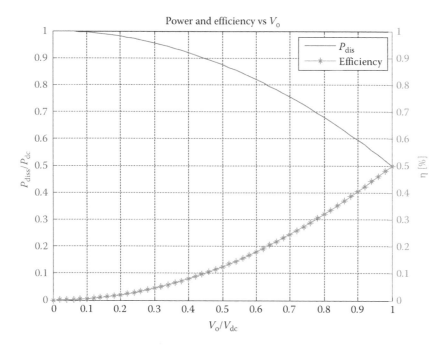

FIGURE 5.20 Class A amplifier power and efficiency diagram.

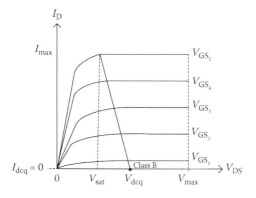

FIGURE 5.21 Typical load line for Class B amplifiers.

$$i_o(\theta) = I_o \sin(\theta) = \frac{m}{n} I \sin(\theta) \qquad (5.50)$$

$$v_o(\theta) = V_o \sin(\theta) = \frac{m}{n} I R_L \sin(\theta) \qquad (5.51)$$

V is the peak value of the transistor voltage, and I is the peak value of the transistor current. The drain-to-source voltages can be written as

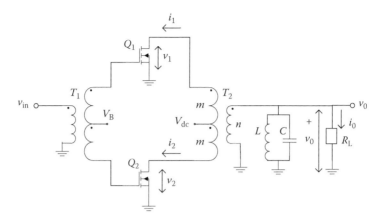

FIGURE 5.22   Practical Class B amplifier in push–pull configuration.

$$v_1(\theta) = V_{dc} + V\sin(\theta) = V_{dc} + \frac{m}{n}V_o\sin(\theta) \qquad (5.52)$$

In one-half of the cycle, only one-half of the primary winding at the output carries current. Therefore, the transformation of the current in m turns of the primary to n turns of the secondary of the output transformer gives the relation for output current to the peak of the transistor voltage as

$$V_o = \frac{n}{m}V \qquad (5.53)$$

$$I_o = \frac{m}{n}I \qquad (5.54)$$

Equation 5.52 can then be written as

$$v_1(\theta) = V_{dc} + V\sin(\theta) = V_{dc} + \left(\frac{m}{n}\right)^2 IR_L\sin(\theta) \qquad (5.55a)$$

or

$$v_1(\theta) = V_{dc} + RI\sin(\theta) \qquad (5.55b)$$

Similarly,

$$v_2(\theta) = V_{dc} - V\sin(\theta) = V_{dc} - IR\sin(\theta) \qquad (5.56a)$$

or

$$v_1(\theta) = V_{dc} + RI\sin(\theta) \tag{5.56b}$$

where

$$R = \left(\frac{m}{n}\right)^2 R_L \tag{5.57}$$

R is the value of the resistance that is seen by one-half of the primary side of the winding. Class B voltage and current waveforms are illustrated in Figure 5.23.

RF output power is

$$P_o = \frac{V_o^2}{2R_L} = \frac{V^2}{2R} \tag{5.58}$$

The maximum output power occurs when $V = V_{dc}$,

$$P_{o,max} = \frac{V_{dc}^2}{2R} \tag{5.59}$$

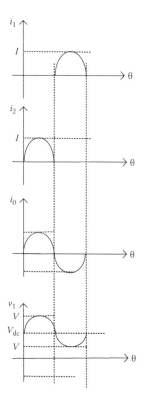

FIGURE 5.23   Class B voltage and current waveforms.

The center tap current can be written as the sum of two drain currents as

$$i_{center,tap}(\theta) = i_1(\theta) + i_2(\theta) = I|\sin(\theta)| \tag{5.60}$$

The DC input current is found from

$$I_{dc} = \frac{1}{2\pi}\int i_{center,tap}(\theta)\, d\theta = \frac{2I}{\pi} = \frac{2}{\pi}\frac{V}{R} \tag{5.61}$$

Therefore, DC input is equal to

$$P_{dc} = V_{dc}I_{dc} = \frac{2}{\pi}\frac{V_{dc}}{R}V \tag{5.62}$$

As a result, the efficiency is equal to

$$\eta_{max} = \frac{P_{out}}{P_{dc}} = \frac{\pi}{4}\frac{V}{V_{dc}} \tag{5.63}$$

The maximum efficiency occurs when $V = V_{dc}$, $\eta_{max} = \frac{\pi}{4}\frac{V}{V_{dc}} \approx 0.7853 \tag{5.64}$

The maximum value of the dissipated power is found when $V = 2V_{dc}/\pi$,

$$P_{d,max} = \frac{2}{\pi^2}\frac{V_{dc}^2}{R} = \frac{4}{\pi^2}P_{o,max} \tag{5.65}$$

5.8   CLASS AB

In Class B mode of operation, amplifier efficiency is sacrificed for linearity. When it is desirable to have an amplifier with better efficiency than Class A amplifier, and yet better linearity than Class B amplifier, then Class AB is chosen as a compromise. The conduction angle for Class AB amplifier is between 180° and 360°. As a result, the bias point for Class AB amplifiers is chosen between the bias points for Class A and Class B amplifiers. Class AB amplifiers are widely used in RF applications when both linearity and efficiency are required. The ideal efficiency of Class AB amplifiers are between 50% and 78.53%. The typical drain-to-source voltage and drain current waveforms are illustrated in Figure 5.24.

5.9   CLASS C

Classes A, B, and AB amplifiers are considered to be linear amplifiers, where the phase and amplitude of output signal are linearly related to amplitude and phase of the input signal. If efficiency is a more important parameter than linearity, then nonlinear amplifier classes such as Class C, D, E, or F can be used. The conduction

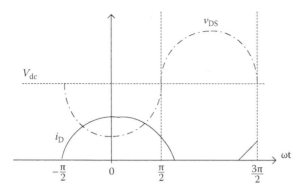

FIGURE 5.24 Typical drain-to-source voltage and drain current for Class AB amplifiers.

angle for Class C amplifier is less than 180°, which makes this amplifier class more efficient than Class B amplifiers.

Class C amplifiers can be analyzed using different models. The models that are known for Class C amplifiers are Current Source Class C Amplifiers, Saturated Class C Amplifiers, and Class C Mixed Mode Amplifiers.

The typical drain-to-source voltage and drain current waveforms for Class C amplifiers are illustrated in Figure 5.25.

The drain efficiency for PA Classes A, B, and C can also be calculated using conduction angle, θ, as

$$\eta = \frac{\theta - \sin\theta}{4\left[\sin(\theta/2) - (\theta/2)\cos(\theta/2)\right]} \tag{5.66}$$

The maximum drain efficiency for Class C amplifier is obtained when $\theta = 0°$. However, the output power decreases very fast when conduction angle approaches and is shown as

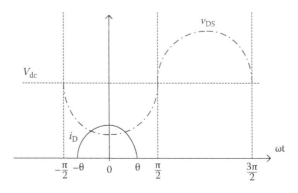

FIGURE 5.25 Typical drain-to-source voltage and drain current for Class C amplifiers.

$$P_o \propto \frac{\theta - \sin(\theta)}{1 - \cos(\theta/2)} \qquad (5.67)$$

As a result, it is not feasible to obtain 100% efficiency with Class C amplifiers. The typical Class C amplifier efficiency in practice is between 75% and 80%.

The efficiency distribution and power capacity of conventional amplifier Classes A, B, AB, and C are illustrated in Figures 5.26 and 5.27, respectively.

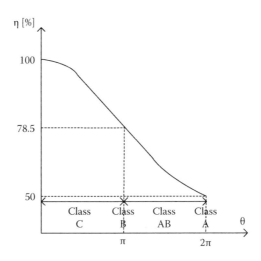

FIGURE 5.26   Efficiency distribution of conventional amplifiers.

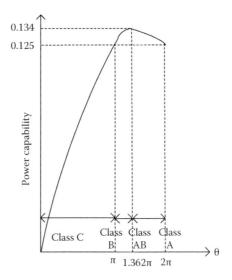

FIGURE 5.27   Power capability of conventional amplifiers.

5.9.1 DESIGN EXAMPLE

Design, simulate, and measure push–pull Class AB amplifier operating at $V_{dc} = 46$ [V] to provide minimum 300 [W] output power using SD1728 50 V epitaxial silicon NPN planar transistor at 27.12 MHz. Then, configure a system to use the same push–pull pair in parallel configuration to obtain minimum power output of 2400 [W]. The transistor data for impedance versus frequency is given in Figure 5.28.

Solution: Class AB amplifier using the method described in this chapter is designed and simulated. The simulated circuit is shown in Figure 5.29.

The prototype has been built using the circuit in Figure 5.29. The simulated and measured results showing gain versus output power are given in Figure 5.30.

As seen from Figure 5.30, power output above 400 was possible with a gain of approximately 12, where measured and simulated results agree.

The measured and simulated collector efficiency versus output power is given in Figure 5.31. The maximum collector efficiency was approximately 60% for 50 Ω.

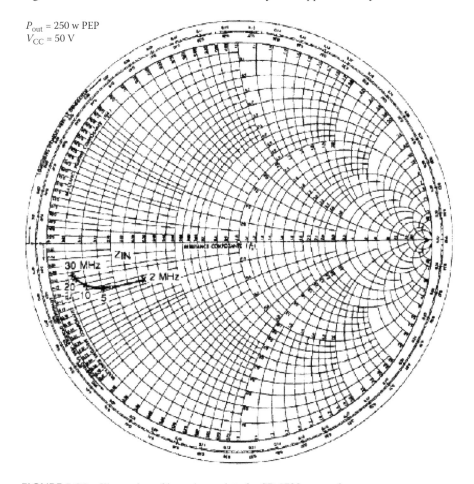

FIGURE 5.28 Illustration of impedance data for SD 1728 versus frequency.

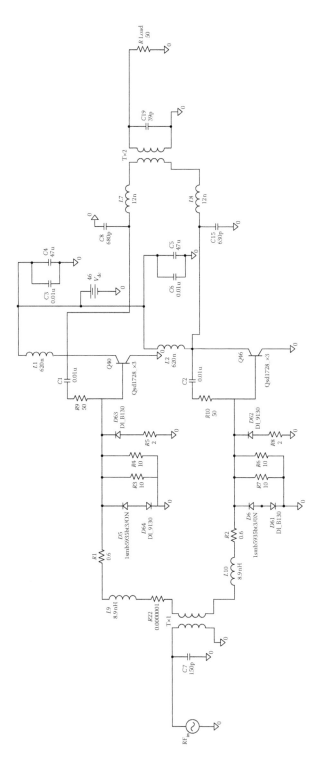

FIGURE 5.29 Simulated Class AB amplifier.

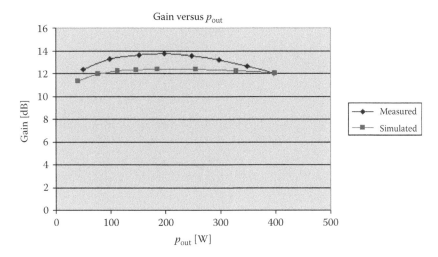

FIGURE 5.30   Class AB amplifier gain versus output power for 50 Ω.

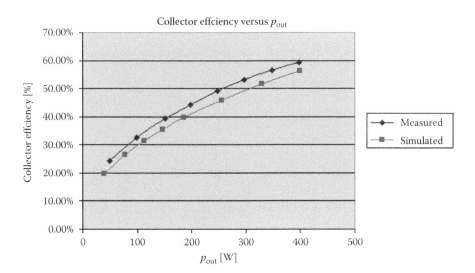

FIGURE 5.31   The simulated and measured results for collector efficiency for 50 Ω.

Then, die dissipation is simulated and measured and illustrated in Figure 5.32. It has been seen that the worst-case die dissipation was approximately 160 [W]. This is well above the dissipation limit of 330 [W] specified for this transistor.

The eight of the push–pull pair is then combined via a combiner to provide a minimum power of 2400 [W] (Figure 5.33). The simulated values for gain versus output power are illustrated in Figure 5.34. It is shown that the system provides above 3200 [W] with 12 dB gain when terminated with 50 Ω.

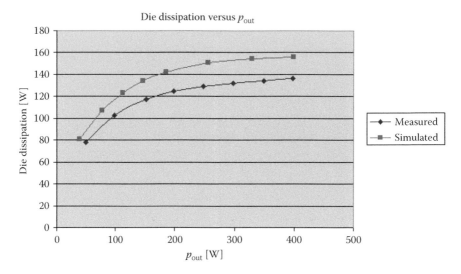

FIGURE 5.32   The simulated and measured results for die dissipation for 50 Ω.

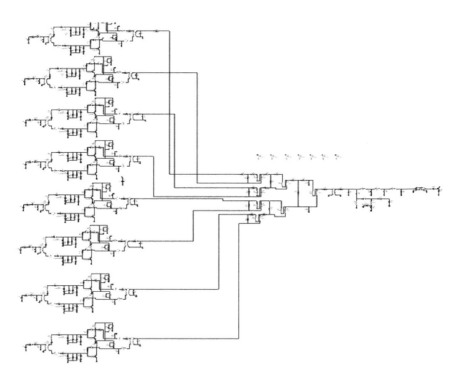

FIGURE 5.33   Illustration of eight push–pull pairs via combiner.

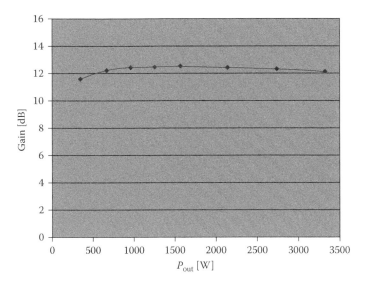

FIGURE 5.34   The gain of combined eight power amplifiers for 50 Ω.

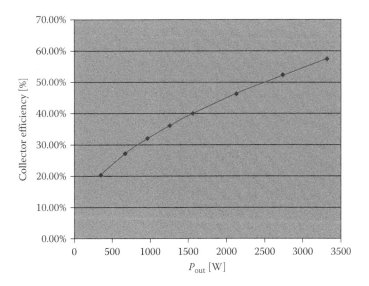

FIGURE 5.35   The efficiency of combined eight power amplifiers for 50 Ω.

The efficiency at 27.12 MHz was approximately 60% for 50 Ω (Figure 5.35). The power contours for 5:1 have been provided via load-pull measurement (Figure 5.36). The die dissipation has been also measured using load-pull measurement (Figure 5.37).

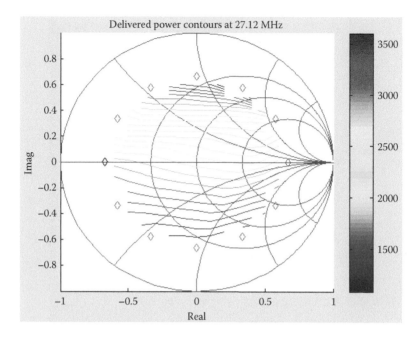

FIGURE 5.36 Power contours for VSWR of 5:1 load using load-pull measurement.

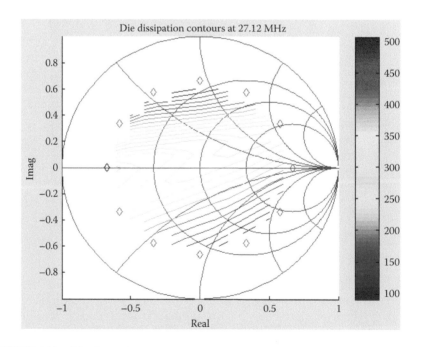

FIGURE 5.37 Die dissipation for VSWR of 5:1 load using load-pull measurement.

REFERENCES

1. J.M. Cusack, S.M. Perlow, B.S. Perlman, Automatic load contour mapping for microwave power transistors. *IEEE Transactions on Microwave Theory Techniques*, Vol. 22, pp. 1146–1152, 1974.
2. G.P. Bava, U. Pisani, V. Pozzolo, Active load technique for load-pull characterisation at microwave frequencies. *Electronic Letters*, Vol. 18, pp. 178–180, 1982.
3. Y. Takayama, A new load-pull characterisation method for microwave power transistors. *IEEE MTT-S International Microwave Symposium Digest*, pp. 218–220, 1976.
4. Williams, T., J. Benedikt, P.J. Tasker, Novel base-band envelope load pull architecture. In *High Frequency Postgraduate Student Colloquium*, pp. 157–161, 2004.
5. M.S. Hashmi, A.L. Clarke, J. Lees, M. Helaoui, P.J. Tasker, F.M. Ghannaouchi, Agile harmonic envelope load-pull system enabling reliable and rapid device characterization, *Measurement Science and Technology*, Vol. 21, p. 055109, 2010.
6. M.A. Pulido, J.A. Reynoso, M.C. Maya, Calibration of a realtime load-pull system using the generalized theory of the TRM technique, *Microwave Measurement Conference, 87th, ARFTG*, pp. 1–4, 2016.
7. M.A. Pulido-Gaytan, J.A. Reynoso-Hernandez, M.C. MayaSanchez, J.R. Loo-Yau, The impact of knowing the impedance of the lines used in the TRL on the load-pull characterization of power transistors, *86th ARFTG Microwave Measurement Conference*, pp. 1–4, 2015.

6 Switch-Mode Amplifier Design and Implementation

6.1 INTRODUCTION

Classes A, AB, and B amplifiers have been used for linear applications where amplitude modulation (AM), single-sideband (SSB) modulation, and quadrature amplitude modulation (QAM) might be required [1–3]. Classes C, D, E, and F are usually implemented for narrowband-tuned amplifiers when high efficiency is desired with high power [4–12]. Classes A, B, AB, and C are operated as transconductance amplifiers, and the mode of operation depends on the conduction angle. In switch-mode amplifiers such as Classes D, DE, E, and F, the active device is intentionally driven into saturation region, and it is operated as a switch rather than a current source. In theory, power dissipation in the transistor can be totally eliminated, and hence, 100% efficiency can be achieved for switch-mode amplifiers.

6.2 CLASS D AMPLIFIERS

Class D amplifiers have two-pole switching operation of transistors either in voltage-mode (VM) configuration that uses series resonator or current-mode (CM) configuration that uses parallel resonator circuit.

6.2.1 VOLTAGE-MODE CLASS D AMPLIFIERS

The two-pole switching operation of transistors in voltage-mode Class D (VMCD) amplifier is illustrated in Figure 6.1. The transfer function of the LCR load network can be found from

$$V_o = IR \tag{6.1a}$$

$$V_{in} = I\left[jwL + \frac{1}{jwC} + R \right] \tag{6.1b}$$

Therefore, the transfer function is

$$H(jw) = \frac{V_o(\omega)}{V_m(\omega)} = \frac{R}{R + j\omega L + \dfrac{1}{j\omega C}} = \frac{R}{R + j\left(\omega L - \dfrac{1}{\omega C} \right)} \tag{6.2}$$

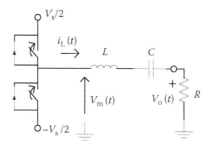

FIGURE 6.1   Class D amplifier in voltage mode.

The natural resonant frequency is then equal to

$$\omega_r = \frac{1}{\sqrt{LC}} \tag{6.3}$$

when $\omega = \omega_r$, $H(j\omega) = 1 = H(j\omega)_{max}$. The quality factor Q and bandwidth are found from

$$Q = \frac{\omega r L}{R} = \frac{1}{\omega_r RC} = \frac{1}{R}\sqrt{\frac{L}{C}} \tag{6.4a}$$

$$BW = \frac{\omega_r}{Q} \tag{6.4b}$$

In Class D operation, it is assumed that the switches are ideal and have zero switching times with no on-resistance or parasitic capacitance. If we assume that the high- and low-side switches are alternately turned *on* and *off* with a 50% duty cycle, $V_m(t)$ can be illustrated as shown in Figure 6.2.

The Fourier Series representation of $V_m(t)$ shown in Figure 6.2 can be written as

$$V_m(t) = a_0 + \sum_{n=1}^{\infty} a_n \cos n\omega_o t + bn \sin n\omega_o t \tag{6.5}$$

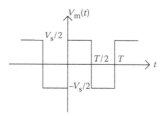

FIGURE 6.2   Waveform for $V_m(t)$.

Fourier coefficients in Equation 6.5 can be obtained from

$$a_{\mathrm{o}} = \frac{1}{T}\int_0^T V_{\mathrm{m}}(t)\,\mathrm{d}t = \frac{1}{T}\left[\int_0^{T/2}\frac{V_{\mathrm{s}}}{2}\,\mathrm{d}t - \int_{T/2}^{T/2}\frac{V_{\mathrm{s}}}{2}\right] = 0 \qquad (6.6\mathrm{a})$$

$$a_n = \frac{2}{T}\int_0^T V_{\mathrm{m}}(t)\cos n\omega_{\mathrm{o}}t\,\mathrm{d}t = \frac{2}{T}\left\{\frac{V_{\mathrm{s}}}{2}\left[\int_0^{T/2}\cos n\omega_{\mathrm{o}}t\,\mathrm{d}t - \int_{T/2}^T\cos n\omega_{\mathrm{o}}t\,\mathrm{d}t\right]\right\}$$

or

$$a_n = \frac{2}{T}\frac{V_{\mathrm{s}}}{2}\left\{\left[\frac{1}{n\omega_{\mathrm{o}}}\sin n\omega_{\mathrm{o}}t\right]_0^{T/2} - \left[\frac{1}{n\omega_{\mathrm{o}}}\sin n\omega_{\mathrm{o}}t\right]_{T/2}^T\right\}$$

$$= \frac{2}{T}\frac{V_{\mathrm{s}}}{2}\left\{\left[\frac{1}{n\omega_{\mathrm{o}}}\sin n\pi\right] - \left[\frac{1}{n\omega_{\mathrm{o}}}\sin 2n\pi\right]\right\} = 0 \qquad (6.6\mathrm{b})$$

Similarly,

$$b_n = \frac{2}{T}\int_0^T V_{\mathrm{m}}(t)\sin n\omega_{\mathrm{o}}t\,\mathrm{d}t \qquad (6.6\mathrm{c})$$

or

$$b_n = \frac{2}{T}\left\{\frac{V_{\mathrm{s}}}{2}\left[\int_0^{T/2}\sin n\omega_{\mathrm{o}}t\,\mathrm{d}t - \int_{T/2}^T\sin n\omega_{\mathrm{o}}t\,\mathrm{d}t\right]\right\}$$

$$= \frac{-2}{T}\frac{V_{\mathrm{s}}}{2}\left\{\left[\frac{1}{n\omega_o}\cos n\omega_{\mathrm{o}}t\right]_0^{T/2} + \left[\frac{-1}{n\omega_{\mathrm{o}}}\cos n\omega_{\mathrm{o}}t\right]_{T/2}^T\right\}$$

which can be expressed as

$$b_n = -\frac{V_{\mathrm{s}}}{T}\left\{\left[\frac{1}{n\omega_{\mathrm{o}}}(\cos n\pi - 1)\right] + \left[\frac{1}{n\omega_{\mathrm{o}}}(-\cos 2n\pi + \cos n\pi)\right]\right\} \qquad (6.6\mathrm{d})$$

Equation 6.6d can be simplified as

$$b_n = -\frac{V_{\mathrm{s}}}{2\pi n}\left[2\cos n\pi - \cos 2n\pi - 1\right] \qquad (6.7\mathrm{a})$$

Hence,

$$b_n = \begin{cases} \dfrac{2V_{\mathrm{s}}}{\pi n} & \text{if } n = 1,3,5,7,\dots \text{ odd} \\ 0 & \text{if } n = 2,4,6,8,\dots \text{ even} \end{cases} \qquad (6.7\mathrm{b})$$

As a result, $V_m(t)$ can be expressed as

$$V_m(t) = \sum_{\substack{n=1,3,5,\dots \\ \text{odd}}}^{\infty} b_n \sin(n\omega_o t) \tag{6.8a}$$

or

$$V_m(t) = \frac{2V_s}{\pi}\sin\omega_s t + \frac{2V_s}{3\pi}\sin 3\omega_s t + \frac{2V_s}{5\pi}\sin 5\omega_s t + \cdots \tag{6.8b}$$

where ω_s is the switching frequency, which is set equal to the resonant frequency, ω_r, of the LCR network to obtain tuned series resonant operation.

We need to find the output voltage across the load resistor. From Equations 6.2 and 6.4,

$$V_o(\omega) = V_m(\omega)H(j\omega) \tag{6.9}$$

From Equation 6.9, it can be seen that the frequency spectrum of the output voltage is the product of $V_m(\omega)$ and $H(j\omega)$.

At the resonant frequency, $\omega = \omega_r$, $H(j\omega) = 1$. Thus, the fundamental component of midpoint voltage will not have any amplitude or phase change. The normalized moduli of the midpoint voltages frequency spectrum, $|V_m(\omega)|$, and the transfer function, $|H(j\omega)|$, are shown in Figure 6.3.

The next significant harmonic, the third harmonic, can be found when $\omega = 3\omega_r$ from Equation 6.2 as

$$H(j3\omega_r) = \frac{R}{R\left[1 + j3\dfrac{\omega_r L}{R} - j\dfrac{1}{3\omega_r RC}\right]} \tag{6.10a}$$

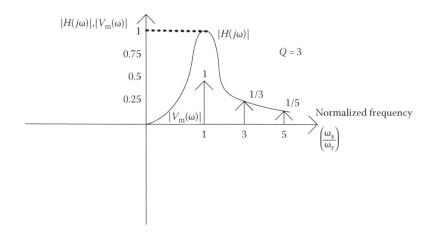

FIGURE 6.3 Frequency spectrum of $|H(j\omega)|, |V_m(\omega)|$ for series LCR network.

which reduces to

$$H(j3\omega_r) = \frac{1}{1+j\frac{8Q}{3}} \quad \text{or} \quad |H(j3\omega_r)| = \frac{3}{\sqrt{9+64Q^2}} \tag{6.10b}$$

From the Fourier Series of the midpoint voltage, $V_m(\omega)$, we know that the amplitude of the third harmonic will be 1/3 of the fundamental. Hence, the ratio of the third harmonic to the fundamental in the output voltage from Equation 6.10 is

$$\frac{|V_{m3}|}{|V_{m_1}|} = \frac{1}{\sqrt{9+64Q^2}} \tag{6.11}$$

For instance, the ratio of the third harmonic to the fundamental is 4% when $Q = 3$. All other harmonics will be reduced considerably less than this.

For the rest of the analysis, we will assume Q to be high enough, so that all harmonics are negligible. Under this assumption, the output voltage can be expressed as

$$V_o(t) = V_{m_1}(t) = \frac{2V_s}{\pi}\sin(\omega_s t) \tag{6.12}$$

The current flowing out of the midpoint to the series LCR network is equal to the load current and expressed as

$$i_L(t) = \frac{V_o(t)}{R} = \frac{2V_s}{\pi R}\sin(\omega_s t) = I_p \sin(\omega_s t) \tag{6.13}$$

where

$$I_p = \frac{2V_s}{\pi R} \tag{6.14}$$

Because duty cycle is 50%, each switch is on half a period and conducts the load current. During the cycle when the switch is *on*, the current reaches the peak value of I_p. Once the switch is turned *off*, the voltage across the switch will be equal to V_s. The waveforms showing the switching operation of Class D amplifier is illustrated in Figure 6.4. The average current $I_{s.avg}$ and the RMS current $I_{s.rms}$ through each switch can then be written as

$$I_{s.avg} = \frac{1}{T_s}\int_0^{T_s/2} I_p \sin(\omega_s t)\,dt = \frac{I_p}{\pi} \tag{6.15}$$

$$I_{s.rms} = \sqrt{\frac{1}{T_s}\int_0^{T_s/2} I_p^2 \sin^2(\omega_s t)\,dt} = \frac{I_p}{2} \tag{6.16}$$

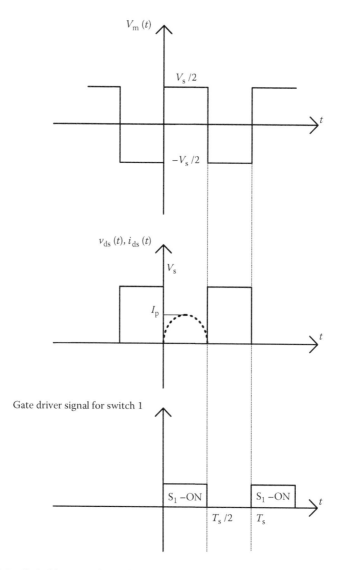

FIGURE 6.4   Switching waveforms for Class D Amplifier—S_1.

Therefore, power at the output is equal to

$$P_{out} = \frac{I_p^2 R}{2} = \frac{2V_s^2}{\pi^2 R} \tag{6.17}$$

which can also be expressed as

$$P_{out} = \frac{V_s I_p}{\pi} \tag{6.18}$$

The power utilization factor, U, which shows the output power capability, can be found for Class D amplifiers from

$$U = \frac{P_{out}}{nV_p I_p} = \frac{1}{2\pi} = 0.159 \tag{6.19}$$

In Equation 6.19, P_{out} represents output, n is the number of switches in the amplifier, and V_p and I_p are the peak voltage and current for each switch, respectively.

In theory, for ideal cases, each switch does not have voltage across it while current is flowing at the same time. This results in an efficiency of 100%. It is important to note that each switch turns on at zero load current and voltage swings from one rail to another.

Having 100% efficiency is impossible in practice because switches are active transistors in real life, and they have nonlinear characteristics and internal parasitics such as output capacitance. When the switch or transistor turns *on*, it has to discharge its output capacitance, whereas output capacitance of the other switch is charged. This results in a capacitive energy loss during transition from *on* to *off*. The total power loss in the switch due to device parasitic capacitance can be expressed as

$$P_D = 2C_o V_s^2 f_s \tag{6.20}$$

In Equation 6.20, f_s is the switching frequency and C_o is the output capacitance of each switch.

6.2.2 CURRENT MODE CLASS D AMPLIFIERS

Current mode Class D (CMCD) amplifier is shown in Figure 6.5. Odd harmonics are terminated by the resonator network in the amplifier. The drain current, $i_d(t)$, can be represented as

$$i_d(t) = I_{dc}\left[\frac{4}{\pi}\sum_{n=1,3,5,\ldots}^{\infty}\frac{1}{n}\sin(n\omega t)\right] \tag{6.21}$$

The peak drain current is equal to

$$I_d = 2I_{dc} \tag{6.22}$$

The output voltage when harmonics are terminated is given as

$$V_o(t) = V_d\sin(\omega t) \tag{6.23}$$

where V_d is the peak drain voltage and is expressed as

$$V_d = \frac{4}{\pi}I_{dc}R_L \tag{6.24}$$

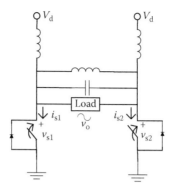

FIGURE 6.5 Current mode Class D amplifier.

The average drain voltage can be found from

$$V_{dd} = \frac{1}{2\pi} \int_{0}^{2\pi} V_d \sin(\omega t) \, d(\omega t) = \frac{1}{\pi} V_d \qquad (6.25)$$

Hence, the peak drain voltage can be written as

$$V_d = \pi V_{dd} \qquad (6.26)$$

From Equations 6.24 and 6.25,

$$I_{dc} = \frac{\pi^2}{4} \frac{V_{dd}}{R_L} \qquad (6.27)$$

The drain current for each switch can be written by considering 180° phase shift between them as

$$i_{d_1}(t) = I_{dc} - \frac{4}{\pi} I_{dc} \sum_{n=1,3,5,\dots}^{\infty} \frac{1}{n} \sin(n\omega t) \qquad (6.28)$$

$$i_{d_1}(t) = I_{dc} + \frac{4}{\pi} I_{dc} \sum_{n=1,3,5,\dots}^{\infty} \frac{1}{n} \sin(n\omega t) \qquad (6.29)$$

The difference between the switch drain currents leads to

$$i_{d_1}(t) - i_{d_2}(t) = -\frac{8}{\pi} I_{dc} \sin(\omega t) - \frac{8}{\pi} I_{dc} \sum_{n=3,5,7,\dots}^{\infty} \frac{1}{n} \sin(n\omega t) \qquad (6.30)$$

When all the odd harmonics are eliminated with the resonator, then Equation 6.30 can be written as

$$i_{d_1}(t) - i_{d_2}(t) = -\frac{8}{\pi} I_{dc} \sin(\omega t) \tag{6.31}$$

The expressions for drain voltages with even harmonics can be obtained similarly as

$$v_{ds_1}(t) = V_{dd} + \frac{\pi}{2} V_{dd} \sin(\omega t) - V_{dd} \sum_{n=1}^{\infty} \frac{2}{(2n)^2 - 1} \cos(2n\omega t) \tag{6.32}$$

$$v_{ds_2}(t) = V_{dd} - \frac{\pi}{2} V_{dd} \sin(\omega t) - V_{dd} \sum_{n=1,2,3,\dots}^{\infty} \frac{2}{(2n)^2 - 1} \cos(2n\omega t) \tag{6.33}$$

The difference between drain-to-source voltage is then equal to

$$v_{ds_1}(t) - v_{ds_2}(t) = \pi V_{dd} \sin(\omega t) \tag{6.34}$$

In Equation 6.34, there are no harmonics, and hence they are being eliminated with this approach, which represents the ideal case. So for ideal case, the impedance at the fundamental frequency is found from

$$Z_{f_o} = \frac{V_{f_o}}{I_{f_o}} = \left| \frac{\pi V_{dd} \sin(\omega t)}{-\frac{8}{\pi} I_{dc} \sin(\omega t)} \right| = \frac{\pi^2}{8} \frac{V_{dd}}{I_{dc}} \quad \text{at} \quad f = f_o \tag{6.35}$$

In addition, the impedance is open, $Z_n = \infty$, for even harmonics, and $Z_n = 0$ for odd harmonics. The summary of the impedance versus harmonics including fundamental is given Table. 6.1.

The waveforms describing the operation of CMCD amplifiers are shown in Figure 6.6.

TABLE 6.1

Summary of Impedance for CMCD Amplifiers for Harmonic Terminations

Frequency—($n\,f_o$)	Impedance—Z_n
f_o	$\dfrac{\pi^2}{8} \dfrac{V_{dd}}{I_{dc}}$
n—Odd	0
n—Even	∞

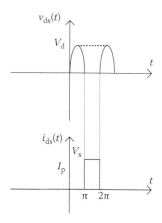

FIGURE 6.6 Current mode Class D amplifier operational waveforms.

6.3 CLASS E AMPLIFIERS

The basic analysis of the Class D amplifiers shows that it is possible to obtain 100% efficiency in theory by modeling the active devices as ideal switches. However, this is not accurate specifically at higher frequencies as device parasitics such as device capacitances play an important role in determining the amplifier performance, which makes the Class D amplifier mode of operation challenging. This challenge can be overcome by making parasitic capacitance of the transistor as part of the tuning network as it is done in Class E amplifiers (Figure 6.7). Class E amplifier shown in Figure 6.7 consists of a single transistor that acts as a switch S, an RF choke, a parallel connected capacitance C_p, a resonator circuit L-C, and a load R_L (Figure 6.8).

6.3.1 CONVENTIONAL ANALYSIS OF CLASS E AMPLIFIERS

In this section, conventional method to design Class E amplifiers for suboptimum and optimum cases will be discussed [4,11].

6.3.1.1 Suboptimum Conditions for Class E Amplifier Design
In the conventional analysis of the circuit shown in Figures 6.7 and 6.8, it is assumed that RF choke is ideal. In the circuit,

- A high Q resonator circuit, LC network, produces sinusoidal output signal at the output of the amplifier.
- The series resonant network L_o and C_o are not tuned at the resonant frequency, f. This creates a reactance that is represented by jX. This can be shown by

$$X = \omega L_o - \frac{1}{\omega C_o} \tag{6.36}$$

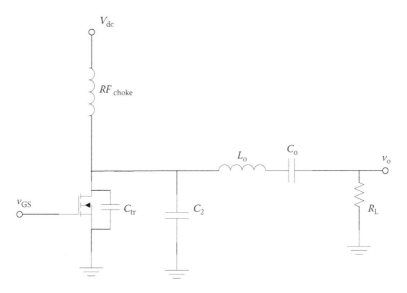

FIGURE 6.7 Simplified circuit of Class E amplifier.

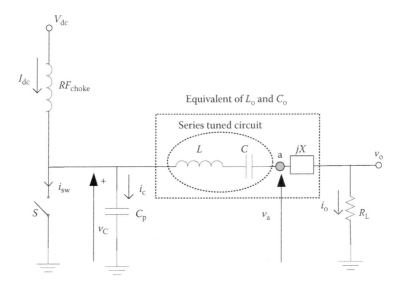

FIGURE 6.8 Equivalent Class E amplifier with switch S.

This effect can be illustrated in the equivalent circuit shown in Figure 6.8. In practice, L and C in Figure 6.8 can be considered to be tuned at the operational frequency, creating a phase shift between the current and the voltage at the input of the tuned circuit. The phase can be used as a knob to adjust for the maximum efficiency.

C_p in the equivalent circuit represents the equivalent capacitance formed by output capacitance of transistor and shunt capacitance in the network. Capacitances are not dependent on voltage.

The circuit analysis can be done with the following conditions [11,13]:

- With the application of the input signal, Switch S is turned *on*, $v_c(\theta) = 0$, $i_c(\theta)$ in capacitor is also zero. The switch current is equal to $i_{sw}(\theta) = I_{dc} - i_o(\theta)$.
- When Switch S is *off*, the switch current $i_{sw}(\theta) = 0$, and the capacitor current is equal to $i_c(\theta) = I_{dc} - i_o(\theta)$. The drain-to-source voltage is created by charging the parallel capacitor.
- When Switch S changes from *off* to *on*, the capacitor discharges.

The waveforms of the voltage and current illustrating the operation of Class E amplifier are given in Figure 6.9. Detailed analysis has been conducted, and waveforms are also given in Reference [4].

In Figure 6.8,

$$v_o(\theta) = V_m \sin(\theta + \phi) \tag{6.37}$$

$$i_o(\theta) = \frac{V_m}{R} \sin(\theta + \phi) \tag{6.38}$$

where $\theta = \omega t$, V_m is magnitude of output voltage, and ϕ is the phase of the output signal. The node voltage, v_a, can be calculated from

$$v_a(\theta) = v_o(\theta) + v_x(\theta) = V_m \sin(\theta + \omega) + \frac{X}{R}\cos(\theta + \omega) = V_a \sin(\theta + \phi_1) \tag{6.39}$$

In Equation 6.39, v_x is defined as the voltage drop across reactance X. The magnitude of the voltage at node a, v_a, is defined as

$$V_a = V_m \sqrt{1 + \left(\frac{X}{R}\right)^2} = V_o \rho \tag{6.40}$$

Its phase

$$\phi_1 = \phi + \psi = \phi + \arctan x \sqrt{\left(\frac{X}{R}\right)} \tag{6.41}$$

The analysis can be performed better by considering the illustration given in Figure 6.10 for *on* and *off* times of the switch.

In Figure 6.10,

–Center of the *off* interval is chosen to be $\pi/2$.
–*Off* length angle is equal to $2y$.

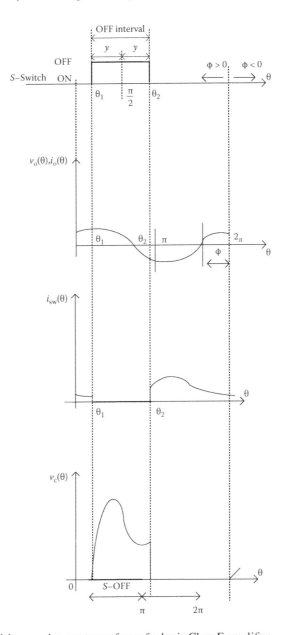

FIGURE 6.9 Voltage and current waveforms for basic Class E amplifier.

Switching instant when switch S is turned *off* is equal to $\theta_1 = \pi/2 - y$ and when $-S$ is turned *on*, it is equal to $\theta_2 = \pi/2 + y$.

When switch is *off*, capacitor will be charging via the current flowing through it. The capacitor voltage is found from

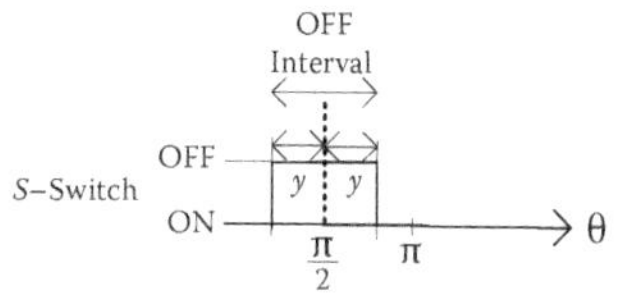

FIGURE 6.10   Illustration of *on* and *off* times of the switch.

$$v_c(\theta) = \frac{1}{\omega C} \int_{\theta_1}^{\theta} i_c(\rho)\,d\rho \tag{6.42}$$

Because the capacitor current, $i_c(\theta)$, $i_c(\theta) = I_{dc} - i_o(\theta)$, $\theta_1 = \pi/2 - y$, Equation 6.42 can be expressed as

$$v_c(\theta) = \frac{1}{\omega C} \int_{\frac{\pi}{2}-y}^{\theta} \left(I_{dc} - i_o(\rho) \right) d\rho \tag{6.43}$$

Substituting Equation 6.38 into Equation 6.43 gives

$$v_c(\theta) = \frac{1}{\omega C_p} \int_{\frac{\pi}{2}-y}^{\theta} \left(I_{dc} - \frac{V_m}{R} \sin(\rho + \phi) \right) d\rho \tag{6.44}$$

Integrating Equation 6.44 gives the capacitor or drain-to-source voltage as

$$v_c(\theta) = \frac{1}{\omega C_p} \left[I_{dc} \left(-\frac{\pi}{2} + y \right) + \frac{V_m}{R} \sin(\phi - y) + I_{dc}\theta + \frac{V_m}{R} \cos(\theta + y) \right] \quad \text{when} \ \ \theta_1 \le \theta \le \theta_2 \tag{6.45a}$$

$$v_c(\theta) = 0 \quad \text{for all other conditions} \tag{6.45b}$$

The series resonant circuit formed by C_o and L_o gets into resonance at the switching frequency, f, and hence has no voltage across them at that frequency. This forces the capacitor voltage, $v_c(\theta)$, to be equal to the node voltage at node a, $v_a(\theta)$. So, the magnitude of $v_a(\theta)$, V_a, can be found from

$$V_a = \frac{1}{\pi} \int_0^{2\pi} v_a(\theta) \sin(\theta + \phi_1)\,d\theta \tag{6.46}$$

The solution of Equation 6.46 is [11]

which leads to

$$V_a = -2\left[\frac{I_{dc}}{\pi\omega C_p}\left(\frac{\pi}{2} - y\right) + \frac{V_m}{\pi\omega C_p R}\sin(y - \phi)\right]\cos\phi_1)\sin(y)$$

$$+ \frac{I_{dc}}{\pi\omega C_p}\left[-2\sin(\phi_1)\sin(y) + \pi\cos(\phi_1)\sin(y) + 2y\sin(\phi_1)\cos(y)\right] \quad (6.47)$$

$$- \frac{V_m}{\pi\omega C_p R}\left[\sin(2\phi + \psi)\sin(2y) - 2y\pi\sin(\psi)\right]$$

Using Equations 6.40 and 6.41, we obtain the magnitude of the output voltage as

$$V_m = I_{dc}\left[\frac{2y\sin(y)\cos(\phi_1) + \sin(\phi_1)\left(2y\cos(y) - 2\sin(y)\right)}{\pi\omega C_p\rho + \frac{1}{2}\sin(2\phi + \psi)\sin(2y) - y\sin(\psi) + 2\sin(y - \phi)\cos(\phi_1)\sin(y)}\right] \quad (6.48)$$

Because the fundamental component of the capacitor voltage does not have a cosine term for ϕ_1 [4], this can be written as

$$0 = \frac{1}{\pi}\int_0^{2\pi} v_c(\theta)\cos\left(\theta + \phi_1\right)d\theta \quad (6.49)$$

Equation 6.49 leads to

$$V_m = I_{dc}R\left[\frac{2y\sin(y)\sin(\phi_1) - 2y\cos(\phi_1)\cos(y) + 2\cos(\phi_1)\sin(y)}{-2\sin(\phi - y)\cos(\phi_1)\sin(y)\sin(\phi_1) - \frac{1}{2}\sin(2y)\cos(2\phi + \psi) + y\cos(\psi)}\right]$$

$$(6.50)$$

Because Equations 6.48 and 6.50 are equal to each other, equating them gives ϕ as

$$\phi = \arctan\left(-\frac{\alpha_2}{\alpha_3}\right) = \arctan\left(-\frac{\alpha_0}{\alpha_1}\right) \quad (6.51)$$

where

$$\alpha_0 = q_0 s_0 - q_1 r_0 \quad (6.52a)$$

$$\alpha_1 = q_1 s_0 + q_0 s_1 + q_0 r_0 - q_1 r_1 \quad (6.52b)$$

$$\alpha_2 = q_1 s_1 + q_0 s_2 + q_0 r_1 - q_1 r_2 \quad (6.52c)$$

$$\alpha_3 = q_1 s_2 + q_0 r_2 \quad (6.52d)$$

and

$$q_0 = 2\big(\sin(y) - 2y\cos(y)\big)\cos(\psi) + 2y\sin(y)\sin(\psi) \tag{6.53a}$$

$$q_1 = 2y\sin(y)\cos(\psi) + 2\big(y\cos(y) - \sin(y)\big)\sin(\psi) \tag{6.53b}$$

$$r_0 = \big(y - \sin(y)\cos(y)\big)\cos(\psi) + 2\sin^2(y)\sin(\psi) \tag{6.54a}$$

$$r_1 = 2\sin^2(y)\sin(\psi) \tag{6.54b}$$

$$r_2 = \big(y - \sin(y)\cos(y)\big)\cos(\psi) \tag{6.54c}$$

$$s_0 = \pi R\omega C_p \rho - \big(y - \sin(y)\cos(y)\big)\sin(\psi) + 2\sin^2(y)\cos(\psi) \tag{6.55a}$$

$$s_1 = -2\sin^2(y)\cos(\psi) \tag{6.55b}$$

$$s_2 = \pi R\omega C_p \rho - \big(y - \sin(y)\cos(y)\big)\sin(\psi) \tag{6.55c}$$

It is important to note that the initial phase angle for ϕ ranges from $-\pi/2$ to $\pi/2$, whereas the solution of Equation 6.51 varies from $-\pi$ to π. Therefore, addition or subtraction of π to the calculated value of ϕ may be needed. The value of the ϕ that should be used is the one when g in the following equation is positive:

$$g = \frac{2y\sin(y)\sin(\phi_1) - 2y\cos(\phi_1)\cos(y) + 2\cos(\phi_1)\sin(y)}{-2\sin(\phi - y)\cos(\phi_1)\sin(y)\sin(\phi_1) - \dfrac{1}{2}\sin(2y)\cos(2\phi + \psi) + y\cos(\psi)} \tag{6.56}$$

The average value of the drain-to-source voltage is equal to DC supply voltage and is found from

$$V_{dc} = \frac{1}{2\pi}\int_0^{2\pi} v_c(\theta)\,d\theta = \frac{I_{dc}}{\pi\omega C_p}\Big[y^2 + yg\sin(\phi - y) - g\sin(\phi)\sin(y)\Big] \tag{6.57}$$

Also,

$$V_{dc} = I_{dc}R_{dc} \tag{6.58}$$

Equating Equations 6.57 and 6.58 gives the value of the DC resistance that is presented to DC supply. Hence, R_{dc} is

$$R_{dc} = \frac{y^2 + yg\sin(\phi - y) - g\sin(\phi)\sin(y)}{\pi\omega C_p} \tag{6.59}$$

The output power of the amplifier in Class E mode is then found using Equations 6.50 and 6.57 as

$$P_o = \frac{1}{2}\frac{V_o^2}{R} = \frac{V_{dc}^2 g^2 R}{2R_{dc}^2} \tag{6.60}$$

The DC input power of the amplifier is found similarly from

$$P_{dc} = V_{dc}R_{dc} = \frac{V_{dc}^2}{R_{dc}} \tag{6.61}$$

Therefore, the drain efficiency of the amplifier is

$$\eta_{drain} = \frac{P_o}{P_{in}} = \frac{g^2}{2}\frac{R}{R_{dc}^2} \tag{6.62}$$

Substituting Equations 6.50 and 6.57 into Equation 6.45 gives the final expression for the drain-to-source voltage as

$$v_c(\theta) = \frac{V_{dc}}{R\omega C_p}\left[\theta + y - \frac{\pi}{2} + g\left(\sin(\phi - y) + \cos(\theta + \phi)\right)\right] \quad \text{when} \quad \theta_1 \leq \theta \leq \theta_2 \tag{6.63a}$$

$$v_c(\theta) = 0 \quad \text{for all other conditions} \tag{6.63b}$$

The maximum value of the drain-to-source voltage is found by taking the derivate of Equation 6.63 with respect to θ as

$$\frac{dv_c(\theta)}{d\theta}\bigg|_{\theta=\theta_{max}} = \frac{V_{dc}}{\omega C_p R_{dc}}\left(1 - g\sin\left(\theta_{max} + \phi\right)\right) = 0 \tag{6.64}$$

Solution of Equation 6.64 for θ_{max} leads to

$$\theta_{max} = \arcsin\left(\frac{1}{g}\right) - \phi \tag{6.65}$$

Substitution of Equation 6.65 into Equation 6.63 gives the maximum value of drain-to-source voltage, $v_c(\theta)$. Similarly, the minimum value of the drain-to-source voltage, $v_c(\theta)$, occurs when

$$\theta_{min} = \pi - \arcsin\left(\frac{1}{g}\right) - \phi \tag{6.66}$$

Earlier, it was given that when the switch was *on*, the switch current was equal to

$$i_{sw}(\theta) = I_{dc} - i_o(\theta) \tag{6.67}$$

Then, the maximum value of the switch current is found as

$$I_{sw,max} = I_{dc} + \frac{V_m}{R} = I_{dc} + \frac{I_{dc}Rg}{R} = I_{dc}(1+g) \tag{6.68}$$

The power utilization factor for Class E amplifiers using Equations 6.60, 6.63, and 6.68 is

$$U = \frac{P_{out}}{V_{c,max} I_{sw,max}} \tag{6.69}$$

6.3.1.2 Optimum Conditions Class E Amplifier Design

In the suboptimum analysis that is closer to practical case without considering any other component loss mechanism, the efficiency degrades due to discharge of capacitor when the switch turns *on*. If the circuit component values are adjusted such that the capacitor voltage drops to zero at the instant when the switch turns *on*, the 100% efficiency is possible for Class E amplifiers. This condition can be met if

$$v_c\left(\frac{\pi}{2} + y\right) = 0 \tag{6.70}$$

From Equations 6.70 and 6.63

$$\cos(\phi) = \frac{y}{g \sin(y)} \tag{6.71}$$

The normalized slope of the capacitor voltage, which is equal to the drain-to-source voltage, is found from

$$\xi = \frac{1}{V_{dc}} \frac{dv_c(\theta)}{d\theta}\bigg|_{\theta = \frac{\pi}{2} + y} = \frac{1}{\omega C_p R_{dc}}\left(1 - g\cos(y + \phi)\right) \tag{6.72}$$

Substituting Equation 6.59 into Equation 6.72 gives

$$\phi_{opt} = \frac{\dfrac{\sin(y)}{y} - \cos(y)}{\dfrac{\xi y}{\pi}\cos(y) - \left(1 + \dfrac{\xi}{\pi}\right)\sin(y)} \tag{6.73}$$

$$g_{opt} = \frac{y}{\cos(\phi_{opt})\sin(y)} \tag{6.74}$$

Because efficiency is 100% for optimum operation, then, $\eta = 1$, and this leads, from Equation 6.59, to

$$R_{dc} = \frac{g_{opt}^2 R}{2} \tag{6.75}$$

In addition, the optimum value of the capacitance, $C_{p,opt}$, is

$$C_{p,opt} = \frac{2\left[y^2 + yg\sin(\phi - y) - g_{opt}\sin(\phi_{opt})\sin(y)\right]}{\pi R \omega g_{opt}^2} \tag{6.76}$$

We can also define the optimum value of ψ, ψ_{opt}, as

$$\psi_{opt} = \arctan\left(\frac{w_1 \sin(\phi_{opt}) + w_2 \cos(\phi_{opt}) + w_3 \cos(2\phi_{opt}) + gy}{w_2 \sin(\phi_{opt}) + w_3 \sin(2\phi_{opt}) - w_1 \cos(\phi_{opt})}\right) \quad (6.77)$$

where

$$w_1 = -2g\sin(\phi_{opt} - y)\sin(y) - 2y\sin(y) \quad (6.78a)$$

$$w_2 = 2y\cos(y) - 2\sin(y) \quad (6.78b)$$

$$w_3 = -g\sin(y)\cos(y) \quad (6.78c)$$

Equations 6.73 through 6.78 give the optimum values of the operational parameters for Class E amplifiers. Any other amplifier parameter for optimum operation can now be obtained using Equations 6.73 through 6.78.

Example 6.1

Design Class E amplifier operating at $f = 1$ MHz, with DC supply voltage of 10 V for suboptimum and optimum cases. Assume $C_p = 700$ [pF], $R = 50$ [Ω], and $Q = 9.92$. Compare results by obtaining the switching waveforms.

Solution: The solution for both suboptimum and optimum cases for Class E amplifier is obtained using Mathcad. The Mathcad algorithm and waveforms are given below (Figures 6.11 through 6.14).
    Suboptimum Case
Given Parameters

$$V_{dc} := 10$$

$$f_0 := 0.987 \cdot 10^6 \quad f := 1 \cdot 10^6 \quad Q_1 := 9.92315 \quad C := 700 \cdot 10^{-12}$$

$$\omega := 2 \cdot \pi \cdot f \quad (\omega_0) := 2 \cdot \pi \cdot f_0 \quad R := 50 \quad y : \frac{\pi}{2}$$

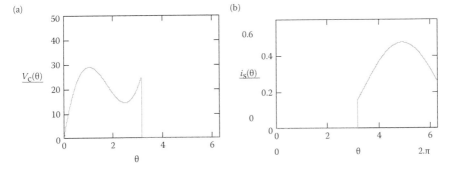

FIGURE 6.11 Suboptimum waveform for Class E amplifier (a) Drain-to-source voltage and (b) switch current.

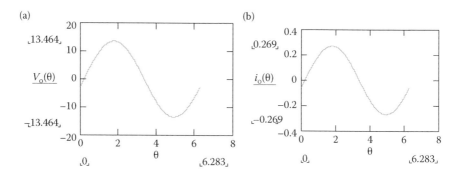

FIGURE 6.12   Suboptimum waveform for Class E amplifier (a) Output voltage and (b) output current.

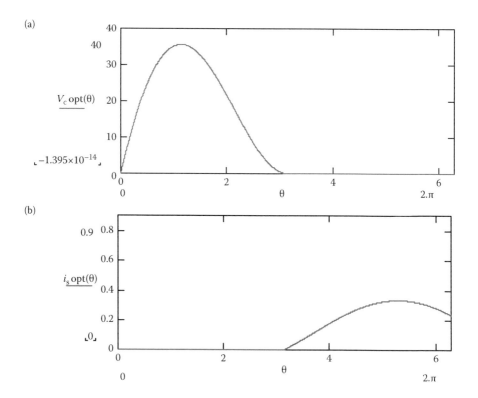

FIGURE 6.13   Optimum waveform for Class E amplifier (a) Drain-to-source voltage and (b) switch current.

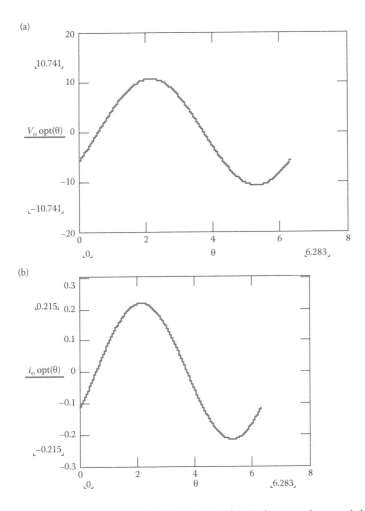

FIGURE 6.14 Optimum waveform for Class E amplifier (a) Output voltage and (b) output current.

Calculation Begins

$$C_0 := \frac{1}{(\omega_0) \cdot R \cdot Q_1} \qquad L_0 := \frac{1}{(\omega_0)^2 \cdot C_0} \qquad B := \omega \cdot C \qquad L_{choke} := 30 \cdot \frac{1}{4 \cdot \pi^2 f^2 \cdot C}$$

$$X := (2 \cdot \pi \cdot f \cdot L_0) - \frac{1}{2 \cdot \pi \, f \cdot C_0}$$

$$\psi := \text{atan}\left(\frac{X}{R}\right)$$

$$\rho := \sqrt{1 + \left(\frac{X}{R}\right)^2}$$

$$q_0 := 2 \cdot \left(\sin(y) - y \cdot \cos(y)\right) \cdot \cos(\psi) + 2 \cdot y \cdot \sin(y) \cdot \sin(\psi)$$

$$q_1 := 2 \cdot y \cdot \sin(y) \cdot \cos(\psi) + 2 \cdot \left[(y) \cdot \cos(y) - \sin(y)\right] \cdot \sin(\psi)$$

$$r_0 := \left(y - \sin(y) \cdot \cos(y)\right) \cdot \cos(\psi) + 2 \cdot \left(\sin(y)\right)^2 \cdot \sin(\psi)$$

$$r_1 := 2 \cdot \left(\sin(y)\right)^2 \cdot \cos(\psi)$$

$$r_2 := \left(y - \sin(y) \cdot \cos(y)\right) \cos(\psi)$$

$$s_0 := \pi \cdot B \cdot R \cdot \rho - \left(y - \sin(y) \cdot \cos(y)\right) \cdot \sin(\psi) + 2 \cdot (\sin(y))^2 \cdot \cos(\psi)$$

$$s_1 := -2 \cdot (\sin(y))^2 \cdot \sin(\psi)$$

$$s_2 := \pi \cdot B \cdot R \cdot \rho - (y - \sin(y) \cdot \cos(y)) \cdot \sin(\psi)$$

$$\alpha_0 := q_0 \cdot s_0 - q_1 \cdot r_0$$

$$\alpha_1 := q_1 \cdot s_0 + q_0 \cdot s_1 + q_0 \cdot r_0 - q_1 \cdot r_1$$

$$\alpha_2 := q_1 \cdot s_1 + q_0 \cdot s_2 + q_0 \cdot r_1 - q_1 \cdot r_2$$

$$\alpha_3 := q_1 \cdot s_2 + q_0 \cdot r_2$$

$$\phi := \text{atan}\left(\frac{-\alpha_0}{\alpha_1}\right)$$

$$\phi_1 := \phi + \psi$$

$$g := \frac{2 \cdot y \cdot \sin(\phi_1) \cdot \sin(y) - 2 \cdot y \cdot \cos(\phi_1) \cdot \cos(y) + 2 \cdot \cos(\phi_1) \cdot \sin(y)}{-2 \cdot \sin(\phi - y) \cdot \sin(y) \cdot \sin(\phi_1) - 0.5 \cdot \sin(2 \cdot y) \cdot \cos(2 \cdot \phi + \psi) + y \cdot \cos(\psi)}$$

$$R_{dc} := \frac{y^2 + g \cdot ((y \cdot \sin(\phi - y) - \sin(\phi) \cdot \sin(y)))}{\pi \cdot B}$$

$$P_o := \frac{(V_{dc})^2 \cdot g^2 \cdot R}{2 \cdot (R_{dc})^2}$$

$$P_{dc} := \frac{(V_{dc})^2}{R_{dc}}$$

$$I_{dc} := \frac{V_{dc} \cdot \pi \cdot B}{\left[\left((y^2 + y \cdot g \cdot \sin(\phi - y) - g \cdot \sin(\phi) \cdot \sin(y))\right)\right]}$$

$$V_1 := I_{dc} \cdot [\cos(\phi_1) \cdot \sin(y) + (2y \cdot \cos(y) - 2 \cdot \sin(y)) \cdot \sin(\phi_1)]$$

$$V_{out} := I_{dc} \cdot R \cdot g$$

$$\theta_{max} := a\sin\left(\frac{1}{g}\right) - \phi$$

$$\theta_{min} := -a\sin\left(\frac{1}{g}\right) - \pi$$

$$I_{s.max} := I_{dc} + \frac{V_{out}}{R}$$

$$\theta_1 := 0$$

$$\theta_2 := \pi$$

$$A(\theta) := \frac{V_{dc}}{B \cdot R_{dc}} \cdot \left[\left(\theta + y - \frac{\pi}{2}\right) + g \cdot (\sin(\phi - y) + \cos(\theta + \phi))\right]$$

$$V_c(\theta) := \text{if}[(\theta_1 \le \theta \le \theta_2), \ A(\theta), \ 0]$$

$$V_o(\theta) := V_{out} \cdot \sin(\theta + \phi)$$

$$i_o(\theta) := \frac{V_{out}}{R} \cdot \sin(\theta + \phi)$$

$$B_s(\theta) := I_{dc} - i_o(\theta)$$

$$i_s(\theta) := \text{if } [(\theta_1 \le \theta \le \theta_2), 0, B_s(\theta)]$$

$$V_{max} := \frac{V_{dc}}{B \cdot R_{dc}} \cdot \left[\left(\theta_{max} + y - \frac{\pi}{2}\right) + g \cdot (\sin(\phi - y) + \cos(\theta_{max} + \phi))\right]$$

$$I_{s.max} := I_{dc} + \frac{V_{out}}{R}$$

$$\text{Capacity} := \frac{P_o}{V_{max} \cdot I_{s.max}}$$

$$\eta := \frac{P_o}{P_{dc}}$$

Power dissipated by discharging the capacitor when switch is turned *off* at

$$P_{dis} := 0.5 \cdot C \cdot (V_c(\pi))^2 \cdot f$$

Given Parameters

$$V_{dc} = 10 \quad C = 7 \times 10^{-10} \quad L_0 = 8.001 \times 10^{-5} \quad C_0 = 3.25 \times 10^{-10} \quad R = 50$$

Calculated Parameters

$$V_{dc} = 10$$

$$X = 12.985 \quad \phi = -0.201 \quad \psi = 0.254 \quad \phi_1 = 0.053 \quad \theta_{max} = 1.05 \quad \theta_{min} = -0.849$$

$$R_{dc} = 49.467 \quad V_{out} = 13.464 \quad I_{dc} = 0.202 \quad I_{s.max} = 0.471 \quad V_{max} = 28.737$$

$$\text{Capacity} = 0.134 \quad P_{dis} = 0.209 \quad P_{dc} = 2.022 \quad \eta = 0.897 \quad P_o = 1.813$$

$$L_{choke} = 1.086 \times 10^{-3} \quad C = 7 \times 10^{-10} \quad C_0 = 3.25 \times 10^{-10} \quad L_0 = 8.001 \times 10^{-5}$$

The waveforms obtained are shown in Figures 6.11 and 6.12.

Optimum Case
The waveforms obtained are shown in Figures 6.13 and 6.14.

6.3.2 HARMONIC MODELING OF CLASS E AMPLIFIERS

An analysis of the Class E power amplifier is given using harmonic modeling that facilitates the characterization of the power amplifier for a wide load-space regime expected in variety of applications. In this section, using a matrix-framework of the direct and quadrature components of dominant harmonics, a quasi-linear algorithm is developed to determine the voltage and current levels expected on all components of the power amplifier for arbitrary loading and operating conditions, with high accuracy [14]. Results of the algorithm are validated by comparison with simulation and experimental results. Applications of the algorithm to circuit design extensions such as phase-controlled Class E power amplifier pairs are described.

Several applications require power amplifier circuit topologies that are conducive to yield higher power conversion density, increased efficiency, and more importantly, enhanced power delivery facilities such as flexible RF envelope waveform generation, while not increasing the size, volume, and heat dissipation profiles. Class E–based power amplifiers are highly suited for high-power density with low-power losses due to single switching stage configuration and the switch-mode operation, and are increasingly used as building blocks in applications such as radio transmitters, plasma generators, etc. However, due to the inherent basis of resonance for power conversion and the switch-mode operation, their control characteristics are nonlinear, which are highly affected by the load impedance and other operating parameters such as the rail voltage, duty cycle of the gate voltage, and the control technique used, such as out-phasing. Moreover, due to the reactive currents and voltages involved in the circuit operation, and the high harmonic content due to switching operation, the voltage and current stresses on critical components are significantly higher than that of conventional "linear" power amplifiers such as Class AB. In addition, the transient and short-term loads expected in practice for some RF applications are far from the standards (such as 50 Ω) the power amplifiers are typically designed or optimized for, and the voltage and current stresses imposed on the power amplifier circuit and components are higher for such nonstandard loads. In view of this, analytical and simulation tools that accurately and quickly characterize the voltage and current profiles of the nonlinear power amplifiers, such as Class E, for arbitrary load and operating conditions are highly desired.

The analysis of single-ended Class E power amplifier using an ideal switch and a high Q-factor for the resonator such that the resultant RF load current is a pure sine wave with no harmonic content is discussed in detail in the literature [4,10,13,15]. The focus of the work reported in these references was to develop the power amplifier design that optimizes it for standard load configurations. For example, in [16], a comprehensive design procedure for the Class E amplifier is given by deriving explicit and accurate design equations and optimizing component selection that shapes the power switch voltage and current waveforms for maximizing power conversion efficiency, assuming a fixed load configuration. An exact analysis of the

Class E amplifier using detailed circuit analysis based on Laplace transforms is presented in [17], which increases the accuracy of design equations taking into account the precise dependency of these equations on the Q-factor. The work reported in Reference [18] extends this analysis to power switch configuration variations, such as that with antiparallel or series diodes, with the objective of improving the efficiency for bounded variations of the load resistor. Reported in [19] is a design procedure for Class E amplifiers, which does not use explicit waveform equations, but instead uses nonlinear algebraic equations based on Poincare mapping of periodical functions describing the amplifier circuit operation and solving them using Newton's iterative algorithm. A noniterative design procedure based on a time-domain analytical solution for the Class E amplifier is presented in Reference [20], which provides increased degrees of freedom in choosing the circuit parameters. A frequency domain-based analysis of the idealized microwave Class E amplifiers is presented in Reference [21], where a generalized harmonic-based model is used to derive a set of linear equations to describe the amplifier functionality.

A major focus of this section is the performance analysis of the Class E–based amplifiers for varied loads, operating conditions, and control strategies, assuming that the design of the amplifier has been optimized for a standard load configuration. The modeling approach taken in this chapter is motivated by the *a priori* qualitative knowledge on the harmonic distortion characteristics of the various voltage and current signals of the Class E amplifier, and each voltage and current waveform can be considered to be composed of multiple, but finite order, harmonics of the RF wave, with the level of harmonic content dependent on the topological location of the voltage or current signal. For example, the voltage across the switching device is rich in second harmonic content, and the input current is rich in DC content, whereas the output voltage and currents typically have low harmonic content due to output filtering. This approach provides an analytical framework made up of algebraic equations that can be solved for the full characterization of the Class E amplifier under varying loads and operating conditions. Although the resulting algebraic equations in general are nonlinear, they are decomposed into a major set of linear equations (corresponding to the passive networks) and a minor set of nonlinear equations (corresponding to the switching networks), which allow the use of simple iterative techniques for solving. The approach presented in this chapter differs from that in Reference [21], where the assumption that the power switch can withstand negative voltage (which ignores the antiparallel diode functionality) simplifies the frequency-domain circuit equations to a linear form.

Figure 6.15 shows the circuit topology of the basic Class E power amplifier stage. It consists of a DC input voltage source with voltage V_{dc}, an input harmonic filter, a single-ended switch S_1 with antiparallel diode D_1, capacitor C_s shunting the switch, a series resonant circuit $L_r - C_r$, and an output harmonic filter feeding into an arbitrary load (shown as an R-C network in the figure, as an example). The power losses for nominal operating load are kept low by minimizing the time-overlap of voltage and current across the switch, which typically is accomplished using zero voltage switching and zero derivative switching. With this, most of the incurred power losses are the result of conduction losses of the switching device.

Analysis of the single-ended Class E power amplifier is performed using the following simplifying assumptions: ideal DC power source, ideal power switch with

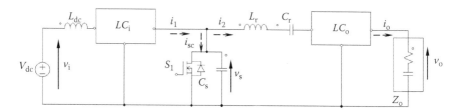

FIGURE 6.15   Circuit configuration of the single-stage Class E power amplifier.

no conduction losses, fixed capacitance across the power switch, resonant inductor–capacitor pair, and the capacitor across power switch optimally chosen for maximum power efficiency for a standard reference load. The analysis is aimed at the full characterization of key circuit voltages and currents of the Class E power amplifier for arbitrary loads and for a variety of control methods such as the duty cycle control, rail voltage control, and the phase control of a power amplifier pair. The circuit variables and parameters of interest include the current and the voltage profile of the power switch, the range and linearity of control achievable with different control methods, and power losses due to any hard-switching of the shunt capacitor voltage.

Referring to Figure 6.15, under steady-state operation, the instantaneous total current i_{sc} (within an RF cycle) through the power switch and the shunt capacitor is written in terms of the current i_1 at the output of the input filter LC_i and the current i_2 through the resonant inductor–capacitor circuit L_r, C_r, using their fundamental and harmonic phasor components (i.e., direct and quadrature axes coordinate values):

$$i_{sc}(\theta) = i_{dc} + \sum_{h=1}^{n} [(i_{1dh} - i_{2dh})\sin(h\theta) + (i_{1qh} - i_{2qh})\cos(h\theta)] \tag{6.79}$$

where θ is an arbitrary angle in radians within an RF cycle, between δ and $2\pi + \delta$, where δ is the phase angle of the RF gate-drive signal with reference to an arbitrary reference; i_{dc} is the DC component of current i_1; h is the harmonic number ranging from 1 to an arbitrary number n; i_{1dh} and i_{2dh} are the direct axis components of the hth harmonic of currents, respectively; i_1 and i_2, and i_{1qh} and i_{2qh} are the quadrature axis components of the hth harmonic of currents, i_1 and i_2, respectively.

During an RF cycle, the average of the current through the power switch S_1, which is turned on at δ and conducts between angles δ and $\psi + \delta$, is given by

$$i_{so} = \frac{1}{2\pi} \int_{\delta}^{\psi+\delta} i_{sc}(\theta) \, d\theta \tag{6.80}$$

In Equation 6.80, $\psi = 2\pi\tau$, where τ is the duty cycle of RF gate control signal. Note that i_{so} should equal i_{dc}, the DC current drawn from the DC input source, and this fact can be used to verify the correctness of any solution that would be obtained through the analysis.

With an antiparallel diode connected across the switch (such as the body diode of a Metal Oxide Field Effect Transistor [MOSFET]), the instantaneous voltage v_s across S_1 at angle q during an RF cycle is either zero or positive, as any potential negative voltage will be clamped to zero by the diode. Typical cases of the switch voltage waveforms are depicted in Figure 6.16. Let us represent the angles at which the positive to zero transitions occur by $\alpha_{2k} = 2\pi + \delta - \varepsilon_{2k}$, and at which the zero to positive transitions occur by $\alpha_{2k+1} = 2\pi + \delta - \varepsilon_{2k+1}, k = 0,1,...., 0 \leq \varepsilon_0 \leq \varepsilon_1 \leq \varepsilon_2 ...$ and ε_0 is such that at $\theta = 2\pi + \delta - \varepsilon_0$, $v_s(\theta)$ reaches zero, or at which the next RF half cycle commences, if $v_s(\theta)$ does not reach zero prior to $2\pi + \delta$. Note that in the latter case, $\varepsilon_0 = 0$.

The positive value of the switch voltage is given by

$$v_s(\theta) = \begin{cases} \dfrac{1}{\omega C_s}\left[\displaystyle\int_{\psi+\delta}^{\theta} i_{sc}(\phi)d\phi\right], & \psi+\delta \leq \theta \leq 2\pi+\delta-\varepsilon_2 \\[3mm] \dfrac{1}{\omega C_s}\left[\displaystyle\int_{2\pi+\delta-\varepsilon_1}^{\theta} i_{sc}(\phi)d\phi\right], & 2\pi+\delta-\varepsilon_1 \leq \theta \leq 2\pi+\delta-\varepsilon_0 \\[3mm] 0, & 2\pi+\delta-\varepsilon_2 \leq \theta \leq 2\pi+\delta-\varepsilon_1 \end{cases} \tag{6.81}$$

In Equation 6.81, w is the RF frequency in radians per second, C_s is the capacitance in parallel with the power switch S_1, and angle φ is the variable for integration. The average voltage v_{s0} across the power switch during an RF cycle is given by

$$v_{s0} = \frac{1}{2\pi}\int_{\psi+\delta}^{2\pi+\delta-\varepsilon_0} v_s(\theta)\, d\theta \tag{6.82}$$

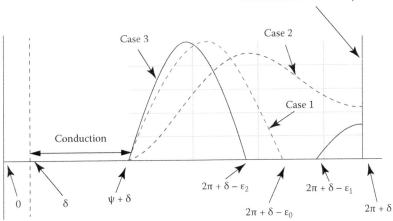

FIGURE 6.16   Description of $\varepsilon_0, \varepsilon_1$ and ε_2 for three typical waveforms of voltage.

Note that v_{so} should equal the DC source voltage V_{dc}.

The direct and quadrature axes components v_{sdh} and v_{sqh}, respectively, of the hth harmonic of the voltage across S_1 are given by

$$v_{sdh} = \int_{\psi+\delta}^{2\pi+\delta-\varepsilon_o} v_s(\theta)\sin(h\theta)\,d\theta \tag{6.83}$$

$$v_{sqh} = \int_{\psi+\delta}^{2\pi+\delta-\varepsilon_o} v_s(\theta)\cos(h\theta)\,d\theta \tag{6.84}$$

The equation relating the direct and quadrature harmonic components of current i_1, the voltage v_s across S_1, and the voltage v_i at the input nodes is given by

$$\begin{pmatrix} v_{idh} \\ v_{iqh} \end{pmatrix} = \begin{pmatrix} k_{adh} & -k_{aqh} \\ k_{aqh} & k_{adh} \end{pmatrix}\begin{pmatrix} v_{sdh} \\ v_{sqh} \end{pmatrix} + \begin{pmatrix} z_{adh} & -z_{aqh} \\ z_{aqh} & z_{adh} \end{pmatrix}\begin{pmatrix} i_{1dh} \\ i_{1qh} \end{pmatrix} \tag{6.85}$$

where scaling factors k_{adh} and k_{aqh} and impedance factors z_{adh} and z_{adh} are constants determined by the input filter network and the DC inductor L_1. Note that with ideal DC voltage source connected at the input node, the direct and quadrature harmonic components v_{idh} and v_{iqh} should be zero.

The equations relating the direct and quadrature harmonic components of current i_2, the voltage v_s across S_1, the output voltage v_o, and the output current i_o are given by,

$$\begin{pmatrix} v_{sdh} \\ v_{sqh} \end{pmatrix} = \begin{pmatrix} k_{bdh} & -k_{bqh} \\ k_{bqh} & k_{bdh} \end{pmatrix}\begin{pmatrix} v_{odh} \\ v_{oqh} \end{pmatrix} + \begin{pmatrix} z_{bdh} & -z_{bqh} \\ z_{bqh} & z_{bdh} \end{pmatrix}\begin{pmatrix} i_{odh} \\ i_{oqh} \end{pmatrix} \tag{6.86}$$

$$\begin{pmatrix} i_{2dh} \\ i_{2qh} \end{pmatrix} = \begin{pmatrix} y_{cdh} & -y_{cqh} \\ y_{cqh} & y_{cdh} \end{pmatrix}\begin{pmatrix} v_{odh} \\ v_{oqh} \end{pmatrix} + \begin{pmatrix} k_{cdh} & -k_{cqh} \\ k_{cqh} & k_{cdh} \end{pmatrix}\begin{pmatrix} i_{odh} \\ i_{oqh} \end{pmatrix} \tag{6.87}$$

where scaling factors k_{bdh}, k_{bqh}, k_{cdh}, and k_{cqh}, impedance factors z_{bdh} and z_{bqh}, and admittance factors y_{cdh} and y_{cqh} are constants determined by the series resonant filter network L_r, C_r and the output filter network LC_o. The equation relating the output current i_o and the output voltages v_o is given by

$$\begin{pmatrix} v_{odh} \\ v_{oqh} \end{pmatrix} = \begin{pmatrix} z_{cdh} & -z_{cqh} \\ z_{cqh} & z_{cdh} \end{pmatrix}\begin{pmatrix} i_{odh} \\ i_{oqh} \end{pmatrix} \tag{6.88}$$

where z_{cdh} and z_{cqh} are the load impedance components.

An exact solution for the single Class E PA configuration can be obtained by solving the system of simultaneous nonlinear equations encompassing expressions 6.82 through 6.88, with v_{idh}, and v_{iqh} set to zero, and v_{s0} set to V_{dc}.

6.4 CLASS DE AMPLIFIERS

The deficiency seen in Class D amplifiers due to capacitive switching losses can be improved by operating the amplifier above the resonant frequency of tuned circuit. This will help reducing the conduction angle of the switches and the duty cycle below 50%.

We begin analysis by assuming the load current to be equal to

$$i_L(t) = I_p \sin(\omega_s t) \tag{6.89}$$

The total change for both output capacitances of the switches is

$$2Q_o = 2C_o V_s \tag{6.90}$$

It can also be written that

$$i_L = \frac{d(2Q_o)}{dt} \tag{6.91}$$

$$d(2Q) = i_L(t)\,dt \tag{6.92}$$

$$\frac{2Q_o}{Q_T} = \int i_L(t)\,dt \tag{6.93}$$

Hence,

$$\int_{\varphi/\omega_s}^{\pi/\omega_s} I_p \sin(\omega_s t)\,dt = 2C_o V_s \tag{6.94a}$$

or

$$\frac{I_p}{\omega_s} - \cos\omega_s t \ \Big|_{\varphi/\omega_s}^{\pi/\omega_s} = 2C_o V_s \tag{6.94b}$$

which leads to

$$\frac{I_p}{\omega_s}\left[-\left(\cos\omega_s \frac{\pi}{\omega_s} - \cos\omega_s \frac{\phi}{\omega_s}\right)\right] = 2C_o V_s \tag{6.95a}$$

or

$$\frac{I_p}{\omega_s}(1 + \cos\phi) = 2C_o V_s \tag{6.95b}$$

Equation 6.95b leads to

$$\cos\phi = \frac{2\omega_s C_o V_s}{I_p} - 1 \quad \text{for } 0 < \phi < \pi \tag{6.96a}$$

So,

$$\phi = \cos^{-1}\left(\frac{2\omega_s C_o V_s}{I_p} - 1\right) \quad \text{for } 0 < \phi < \pi \tag{6.96b}$$

where φ is the conduction angle. The conduction angle φ is between $0°$ and $180°$.

Using Equation 6.96, the maximum frequency of operation is obtained when $\varphi = 0$ as

$$1 = \frac{2\omega_s C_o V_s}{I_p} - 1 \tag{6.97a}$$

or

$$f_s = \frac{I_p}{2\pi C_o V_s} \tag{6.97b}$$

The average switch current can be found from

$$I_{s.avg} = \frac{1}{T_s} \int_0^{t_{ON}} I_p \sin(\omega_s t)\, dt \tag{6.98a}$$

or

$$I_{s.avg} = \frac{I_p}{2\pi}(1 - \cos\phi) \tag{6.98b}$$

In addition, we can calculate the rms value of the switch current as

$$I_{s.rms} = \sqrt{\frac{1}{T_s} \int_0^{t_{ON}} I_p^2 \sin^2(\omega_s t)\, dt} \tag{6.99a}$$

or

$$I_{s.rms} = \frac{I_p}{2}\sqrt{\frac{(2\varphi - \sin 2\phi)}{2\pi}} \tag{6.99b}$$

Apart from the usual ratings of a switch, another specification is of importance at high frequencies and pertains to the allowed rate of change of voltage across the switch. The maximum value of the rate of change for voltage (d_v/d_t) imposed on a switch occurs when the switch turns *off*. When the switch turns *off*,

$$\frac{d_v}{d_t} = \frac{I}{C} \tag{6.100}$$

where $i_L(t) = I_p\sin\phi$. So, Equation 6.100 can be written as

$$\frac{dV}{dt}\bigg|_{max} = \frac{I_p}{2c_o}\sin\phi \tag{6.101}$$

The output power is found from,

$$P_{out} = \frac{V_s I_p}{2\pi}(1 - \cos\phi) \tag{6.102}$$

Power output depends on conduction angle and reaches its maximum value when ϕ is 180°.

The power utilization factor for Class DE operation is found from

$$U = \frac{P_{out}}{nV_p I_p} = \frac{V_s I_p}{n(V_p I_p)2\pi}(1 - \cos\phi) \tag{6.103}$$

Since the number of switches is 2, $n = 2$, then U is equal to

$$U = \frac{(1 - \cos\phi)}{4\pi} \tag{6.104}$$

U is again maximum and is equal to $U = 0.159$ when $\phi = 180°$. This shows that the operation of Class DE amplifier remains in Class D operation when conduction angle is $\phi = 180°$.

$V_m(t)$ in Figure 6.17 during $t_{on} < t < T_s/2$ can be found from

$$V_m(t) = -\frac{1}{2c_o}\int I_p\sin(\omega_s t)\, dt + K \tag{6.105}$$

Since $V_m(t_{ON}) = \frac{V_s}{2}$, the solution of Equation 6.105 can be written as

$$V_m(t) = \frac{V_s}{(1 + \cos\phi)}\cos(\omega_s t) + \frac{V_s}{2}\left(\frac{1 - \cos\phi}{1 + \cos\phi}\right) \quad \text{for } t_{on} < t < T_s/2 \tag{6.106}$$

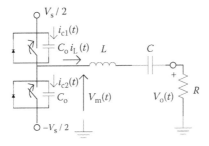

FIGURE 6.17 Class DE amplifier with each switch having on output capacitance c_o.

The fundamental component of $V_m(t)$ can be found using Fourier analysis as

$$v_{m_1}(t) = 2\frac{V_s}{2\pi(1+\cos\phi)}\sqrt{(\pi-\phi\sin\phi\cos\phi)^2+\sin^4\phi}\cos(\omega_s t+\beta) \quad (6.107)$$

where

$$\beta = \tan^{-1}\left[\frac{-\sin^2\phi}{\pi-\phi+\sin\phi\cos\phi}\right] \quad (6.108)$$

Substitution of maximum and minimum values of ϕ into Equation 6.107 gives the magnitude of the fundamental component as

$$V_{m_1} = \frac{2V_s}{\pi} \quad \text{for } \phi = 180° \quad (6.109a)$$

$$V_{m_1} = \frac{V_s}{2} \quad \text{for } \phi = 180° \quad (6.109b)$$

We introduce the phase angle, α, between the fundamental and the load current as

$$\alpha = \beta + \frac{\pi}{2} \quad (6.110)$$

Substituting Equation 6.108 into Equation 6.110 gives

$$\alpha = \tan^{-1}\left[\frac{\pi-\phi+\sin\phi\cos\phi}{\sin^2\phi}\right] \quad (6.111)$$

It is seen from Equation 6.111 that Class D operation is obtained when $\varphi = 180°$, $\alpha = 0°$. At this phase angle, load current will be in phase with the fundamental. On the other hand, when $\varphi = 0°$, $\alpha = 90°$, load will be an inductive load.

6.4.1 ANALYSIS OF SERIES RESONANT NETWORK

Consider the series resonant LCR network in Figure 6.17, which is given as a reference in Figure 6.18.

The resonant frequency of the network shown in Figure 6.18 is

$$\omega_r = \frac{1}{\sqrt{LC}} \quad (6.112)$$

We can define the loaded quality factor for the network as

$$Q = \frac{\omega_r L}{R} = \frac{1}{\omega_r CR} \quad (6.113)$$

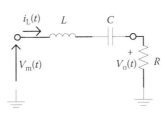

FIGURE 6.18 Series resonant LCR network.

The equivalent impedance of the network is obtained as

$$Z = R + j\left(\omega L - \frac{1}{\omega C}\right) = R\left[1 + jQ\left(\frac{\omega}{\omega_r} - \frac{\omega_r}{\omega}\right)\right] \qquad (6.114)$$

Z in Equation 6.114 is a complex impedance and can also be written as

$$Z = R + jX = |Z|\angle\theta^\circ \qquad (6.115)$$

where

$$|Z| = R\sqrt{\left[1 + jQ^2\left(\frac{\omega}{\omega_r} - \frac{\omega_r}{\omega}\right)^2\right]} \qquad (6.116a)$$

$$\theta = \tan^{-1}\left[Q\left(\frac{\omega}{\omega_r} - \frac{\omega_r}{\omega}\right)\right] \qquad (6.116b)$$

and

$$R = |Z|\cos\theta \qquad (6.117a)$$

$$X = |Z|\sin\theta \qquad (6.117b)$$

If the frequency of operation for the series resonant LCR network is above the reso-nant frequency, then load current will lag the fundamental component of $v_m(t)$. This is a requirement for the amplifier to operate in Class DE mode. This can be better understood by considering the modified circuit illustrating the operation above reso-nant frequency in Figure 6.19.

When the network shown in Figure 6.19 gets into resonance, the fundamental component of $v_m(t)$ will see impedance, Z', which is equal to

$$Z' = R + jX' \qquad (6.118)$$

The new impedance, Z', can be found using circuit analysis as

$$Z' = \frac{V_s}{\pi I_p}(1 - \cos\phi) + j\frac{V_s}{\pi I_p}\left(\frac{\pi - \phi + \sin\phi\cos\phi}{1 + \cos\phi}\right) \qquad (6.119)$$

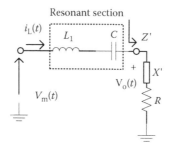

FIGURE 6.19　Modified circuit for operation above resonant frequency.

From Equations 6.118 and 6.119,

$$R = \frac{V_s}{\pi I_p}(1 - \cos\phi) \tag{6.120}$$

$$X' = \frac{V_s}{\pi I_p}\left(\frac{\pi - \phi + \sin\phi\cos\phi}{1 + \cos\phi}\right) \tag{6.121}$$

when $\dfrac{V_s}{I_p} = 1$, R', and X' can be plotted versus ϕ.

　　The component values for Class DE operation are found from Equation 6.120 and

$$\tan\theta = \tan\alpha = \frac{X}{R} = \frac{\pi - \phi + \sin\phi\cos\phi}{\sin^2\phi} = Q\left(\frac{\omega}{\omega_r} - \frac{\omega_r}{\omega}\right) \tag{6.122}$$

where $\omega = \omega_s$ in Equation 6.122. From Equation 6.122, we obtain the resonant frequency as

$$\omega_r = \frac{\omega}{2}\left(\sqrt{\left(\frac{\tan\alpha}{Q}\right)^2 + 4} - \frac{\tan\alpha}{Q}\right) \tag{6.123a}$$

or

$$f_r = \frac{f}{2}\left(\sqrt{\left(\frac{\tan\alpha}{Q}\right)^2 + 4} - \frac{\tan\alpha}{Q}\right) \tag{6.123b}$$

Q is the designer's choice and will be selected between 3 and 5 for Class DE operation. Once ω_r is known, L and C in the original series resonant network given in Figure 6.18 are calculated from

$$L = \frac{QR}{\omega_r} \tag{6.124a}$$

$$C = \frac{1}{\omega_r^2 L} \tag{6.124b}$$

The summary of the design procedure for Class DE amplifier can be outlined as follows:

- Design parameters including operating voltage, frequency, peak current, C_{oss}, and V_{ds} are given.
- Calculate conduction angle, ϕ, from Equation 6.96b.

$$\phi = \cos^{-1}\left(\frac{2\omega_s C_o V_s}{I_p} - 1\right) \tag{6.125}$$

If there is large variation in the output capacitance of the transistor, C_{oss}, then use data sheet information for Q_T vs V_{ds}. For instance, the typical data sheet showing C_{oss} versus V_{ds} for MOSFET transistor is given in Figure 6.20.

There is a large variation in C_{oss} as shown in Figure 6.20. Hence, using C_{oss} given might not lead to an accurate conduction angle, ϕ. Instead, it might be better to use charge information for the transistor. When charge information is available for the transistor, we can use the following equation for the conduction angle:

$$\phi = \cos^{-1}\left(\frac{2\omega_s Q_{Ts}}{I_p} - 1\right) \tag{6.126}$$

where Q is the total charge needed to raise V_{ds} of transistor to the given supply voltage, V_s.

$$Q_T = C_o V \tag{6.127}$$

Calculate P_o from Equation 6.102.

$$P_{out} = \frac{V_s I_p}{2\pi}(1 - \cos\phi) \tag{6.128}$$

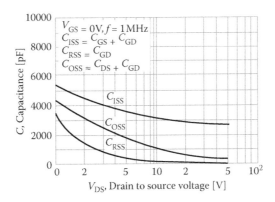

FIGURE 6.20 Typical MOSFET C_{oss} versus V_{ds} given in manufacturer data sheet. (From IRF450 Transistor Datasheet. http://www.irf.com/product-info/datasheets/data/jantx2n6770.pdf.)

Find phase angle, α, between load current and fundamental of v_m (t) from Equation 6.111.

$$\alpha = \tan^{-1}\left[\frac{\pi - \phi + \sin\phi\cos\phi}{\sin^2\phi}\right] \tag{6.129}$$

Choose Q between 3 and 5.

Calculate the resonant frequency of the series resonant network from Equation 6.123b.

$$f_r = \frac{f}{2}\left(\sqrt{\left(\frac{\tan\alpha}{Q}\right)^2 + 4} - \frac{\tan\alpha}{Q}\right) \tag{6.130}$$

Calculate the component values R, L, and C of the resonant network from Equations 6.124 and 6.120 as

$$L = \frac{QR}{\omega_r} \tag{6.131a}$$

$$C = \frac{1}{\omega_r^2 L} \tag{6.131b}$$

$$R = \frac{V_s}{\pi\,I_p}(1 - \cos\phi) \tag{6.131c}$$

Find the average and rms values of the switch current from Equation 6.99 as

$$I_{s.rms} = \sqrt{\frac{1}{T_s}\int_0^{t_{ON}} I_p^2\sin^2(\omega_s t)\,dt} \tag{6.132a}$$

$$I_{s.rms} = \frac{I_p}{2}\sqrt{\frac{(2\phi - \sin 2\phi)}{2\pi}} \tag{6.132b}$$

Example 6.2

Assume that the following operational parameters for the amplifier are given:

Operating voltage: $V_s = 300$ V

Peak current: $I_p = 16$ A

Operating frequency: $f = 5$ mHz

Transistor Parameters: Rated Voltage: 500 V, Rated Current : 14 A. Co = 370 pF. Design this amplifier to operate in Class DE mode for loaded Q of 3.75.

Solution: From the given data, the conduction angle, ϕ,

$$\phi = \cos^{-1}\left(\frac{2\omega_s C_o V_s}{I_p} - 1\right) = 124.54°$$

The output power is found as

$$P_o = \frac{V_s I_p}{2\pi}(1 - \cos\varphi) = 1195 \text{ [W]}$$

The phase angle, α, between the fundamental component of the midpoint voltage and the load current is equal to

$$\alpha = \tan^{-1}\left[\frac{\pi - \phi + \sin\phi\cos\phi}{\sin^2\phi}\right] = 36.6°$$

The resonant frequency is found as

$$f_r = \frac{f_s}{2}\left(\sqrt{\left(\frac{\tan\alpha}{Q}\right)^2 + 4} - \frac{\tan\alpha}{Q}\right) = 4.53 \text{ [MHz]}$$

The design component values R, L, and C are found as

$$R = \frac{V_s}{\pi I_p}(1 - \cos\phi) = 9.34$$

$$L = \frac{QR}{\omega_r} = 1.23 \text{ μH}$$

and

$$C = \frac{1}{\omega_r^2 L} = \frac{1}{\omega_r QR} = 1.006 \text{ nF}$$

The average and rms values of the switch current are calculated as

$$I_{s.avg} = \frac{I_p}{2\pi}(1 - \cos\phi) = 3.98 \text{ [A]}$$

$$I_{s.rms} = \frac{I_p}{2}\sqrt{\frac{(2\phi - \sin 2\phi)}{2\pi}} = 7.33 \text{ [A]}$$

6.5 CLASS F AMPLIFIERS

Class F amplifiers carry similar characteristics to Class B or Class C amplifiers that use a single-resonant load network to produce simple sinusoid at the resonant frequency. The power capacity and efficiency obtained for Class B and Class C amplifiers can be improved by introducing the harmonic terminations of the load network as it is done in Class F. Therefore, the load network in Class F amplifier resonates at the operational frequency as well as one or more harmonic frequencies. Multiresonant load network in Class F amplifier helps controlling the harmonic contents of the drain voltage and current, and wave shapes to minimize the overlap region between them to reduce the transistor power dissipation. This results in improvement of both efficiency and power capacity. In theory, an ideal Class F amplifier can control an infinite number of harmonics, has a square voltage waveform, and can give 100% efficiency. However, in practical applications, it is very difficult to control above the fifth harmonic with Class F amplifiers. The schematics of the basic Class F amplifier with the third harmonic, called Class F$_3$, peaking and its voltage and current waveforms are given in Figure 6.21a and b, respectively. Class F amplifier shown in Figure 6.21 has two parallel LC resonators, which are tuned to center frequency, f_o, and third harmonic frequency, $3f_o$.

Consider the general Class F amplifier circuit given in Figure 6.22 for the analysis as given in Reference [23].

The drain voltage and current waveforms are expressed as

$$v_d(\theta) = V_{dd} + V_{0m}\sin(\theta) + V_{3m}\sin(3\theta) + V_{5m}\sin(5\theta) + \cdots \tag{6.133}$$

$$i_d(\theta) = I_{dd} - I_{0m}\sin(\theta) + I_{2m}\cos(2\theta) - I_{4m}\cos(4\theta) + \cdots \tag{6.134}$$

where $\theta = \omega t$. The coefficients of the fundamental component can be related to DC component and peak of the waveform as:

$$V_{0m} = \gamma_V V_{dd} \tag{6.135a}$$

$$v_{d\max} = \delta_V V_{dd} \tag{6.135b}$$

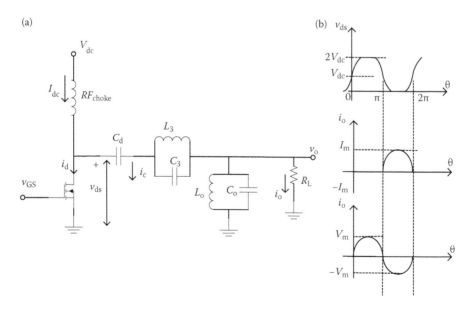

FIGURE 6.21 (a) Class F amplifier with third harmonic peaking and (b) Voltage and current waveforms for Class F amplifier with third harmonic peaking.

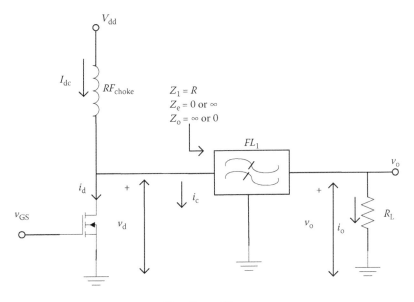

FIGURE 6.22 Generic circuit for Class F amplifier.

$$I_{0m} = \gamma_1 I_{dc} \tag{6.136a}$$

$$i_{d\max} = \delta_1 V_{dd} \tag{6.136b}$$

It is shown in Reference [23] that the load presents drain at the operation frequency

$$Z(f) = R \tag{6.137}$$

The output voltage and current are defined by

$$V_{0m} = I_{0m} R \tag{6.138}$$

Power output is then found as

$$P_o = \frac{V_{0m}^2}{2R} = \frac{\gamma_v^2 V_{dd}^2}{2R} \tag{6.139}$$

Similarly, DC input current and DC input power can be written as

$$I_{dc} = \frac{I_{0m}}{\gamma_1} = \frac{V_{0m}}{R\gamma_1} = \frac{\gamma_v V_{dd}}{R\gamma_1} \tag{6.140}$$

$$P_i = V_{dd} I_{dc} = \frac{\gamma_v V_{dd}^2}{\gamma_1 R} \tag{6.141}$$

The efficiency of the amplifier is calculated from

$$\eta = \frac{P_o}{P_i} = V_{dd}I_{dc} = \frac{\gamma_V \gamma_1}{2} \tag{6.142}$$

Because the voltage and current relations are defined by Equations 6.15 through 6.136, harmonic coefficients of a maximally flat waveform can then be varied to have the maximum and/or minimum values of the voltage or current waveforms. The maximum and minimum values of the voltage waveform occur at $\theta = \pi/2$ and $3\pi/2$. These values can be substituted into the derivatives and solve for the basic waveform parameters for the odd harmonic voltage wave. This can be illustrated on the voltage waveform by taking the derivative of Equation 6.133. The odd order derivatives are zero since $\cos(\pi/2) = \cos(3\pi/2) = 0$. The maximum flatness at the minimum value of the voltage requires the even derivatives to be zero when $\theta = 3\pi/2$. So, from Equation 6.133,

$$\frac{d^2 v_d}{d\theta^2} = -V_{0m}\sin(\theta) - 9V_{3m}\sin(3\theta) - 25V_{5m}\sin(5\theta) \tag{6.143}$$

So,

$$\left.\frac{d^2 v_d}{d\theta^2}\right|_{\theta=\frac{3\pi}{2}} = V_{0m} - 9V_{3m} + 25V_{5m} \tag{6.144}$$

Similarly,

$$\frac{d^4 v_d}{d\theta^4} = V_{0m}\sin(\theta) + 81V_{3m}\sin(3\theta) + 625V_{5m}\sin(5\theta) \tag{6.145}$$

So,

$$\left.\frac{d^4 v_d}{d\theta^4}\right|_{\theta=\frac{3\pi}{2}} = -V_{0m} + 81V_{3m} - 625V_{5m} \tag{6.146}$$

Third Harmonic Peaking

The third harmonic peaking occurs when $V_{5m} = 0$. Substitution of this value into Equation 6.144 gives

$$V_{3m} = \frac{1}{9}V_{0m} \quad \text{when} \quad V_{5m} = 0 \tag{6.147}$$

When $\theta = 3\pi/2$, Equation 6.133 is equal to

$$0 = V_{dd} - V_{0m} + V_{3m} \tag{6.148}$$

From Equations 6.147 and 6.148,

$$V_{0m} = \frac{9}{8}V_{dd} \tag{6.149}$$

$$V_{3m} = \frac{1}{8}V_{dd} \tag{6.150}$$

The peak value of the drain voltage is found as

$$v_{d\,max} = V_{dd} + V_{0m} - V_{3m} = 2V_{dd} \tag{6.151}$$

Hence, the voltage coefficients in Equations 6.135 and 6.136 found for third harmonic peak are

$$\gamma_V = \frac{9}{8} \tag{6.152a}$$

and

$$\delta_V = 2 \tag{6.152b}$$

Fifth Harmonic Peaking
Solution of Equations 6.144 and 6.146 for V_{3m} and V_{3m} gives

$$V_{3m} = \frac{1}{6}V_{0m} \quad \text{and} \quad V_{5m} = \frac{1}{50}V_{0m} \tag{6.153}$$

When $\theta = 3\pi/2$, Equation 6.133 is equal to

$$0 = V_{dd} - V_{0m} + V_{3m} - V_{5m} \tag{6.154}$$

From Equations 6.153 and 6.154,

$$V_{0m} = \frac{75}{64}V_{dd} \tag{6.155}$$

$$V_{3m} = \frac{25}{128}V_{dd} \tag{6.156}$$

$$V_{5m} = \frac{3}{128}V_{dd} \tag{6.157}$$

The peak value of the drain voltage is found from

$$v_{d\,max} = V_{dd} + V_{0m} - V_{3m} + V_{5m} = 2V_{dd} \tag{6.158}$$

Hence, the voltage coefficients in Equations 6.135 and 6.136 found for the fifth harmonic peak are

$$\gamma_V = \frac{75}{64} \tag{6.159a}$$

and

$$\delta_V = 2 \tag{6.159b}$$

The results of similar analysis applied for odd and even harmonics are tabulated in Tables 6.2 and 6.3 for voltage and current coefficients, respectively.

The maximum efficiency and power output of Class F power amplifiers with different numbers of harmonics are tabulated in Table 6.4.

The component values for Class F amplifier are obtained as

$$R = \frac{\gamma_V^2 V_{dd}^2}{2P_o} \tag{6.160}$$

The resonant circuit capacitor values are obtained from

$$C_i = \frac{1}{\left(2\pi f\right)^2 L_i} \tag{6.161}$$

TABLE 6.2
Voltage Waveform Coefficients

Desired Harmonics	δ_V	$\gamma_V = V_{0m}/V_{dd}$	V_{3m}/V_{0m}	V_{5m}/V_{0m}
1	2	1	—	—
3	2	1.1547	0.1667	—
5	2	1.0515	—	−0.0618
3 + 5	2	1.2071	0.2323	0.0607
∞	2	1.273	0.424	0.255

TABLE 6.3
Current Waveform Coefficients

Desired Harmonics	δ_I	$\gamma_I = I_{0m}/I_{dc}$	I_{2m}/I_{0m}	I_{4m}/V_{dd}
1	2	1	—	—
2	2.9142	1.4142	0.3540	—
4	2.1863	1.0824	—	−0.0957
2 + 4	3	1.5	0.3890	0.0556
∞	3.142	1.571	0.667	0.133

TABLE 6.4

Maximum Efficiency and Power Output of Class F Power Amplifiers

Max Harmonic Current	Efficiency, η			
m	Max Harmonic-Voltage, $n = 1$	Max Harmonic-Voltage, $n = 3$	Max Harmonic-Voltage, $n = 5$	Max Harmonic-Voltage, $n = \infty$
1	0.5	0.5774	0.6033	0.6370
2	0.7071	0.8165	0.8532	0.9003
4	0.7497	0.8656	0.9045	0.9545
∞	0.785	0.9069	0.9477	1
$P_{o,max}$	0.125	0.1443	0.1508	0.159

The following Matlab program has been developed to design any third- or fifth-order peaking Class F amplifier and is used for design of these types of amplifiers:

```
% Class F Amplifier Design Program for 3rd and 5th Harmonic
Peaking
clear;
clc;
%PA design parameters
harmonics = 0;
%Selection of Harmoni Design
while (harmonics *= 1 && harmonics *= 2)
harmonics= input( 'Please Type 1 for 3th-harmonic design or 2
for 5th-harmonic design : '  );
end


fo = input( 'Please Enter the frequency of operation, fo
(MHz): ');
if (fo <= 0)
    fo=1e9; %Frequency of operation
else
fo=fo*1e6;
end

w = 2*pi*fo;
Po= input( 'Please enter the desired output power in mW, Po
(mW): ');

Vdc= input( 'Please enter the value of the regulated power
supply Vdc: ');

L0 = input( 'Please enter the inductance for the base filter
(nH): ');
```

```
if (L0 <= 0)
    L0=1e-9; %Base filter inductance
else
L0=L0 * 1e-9;
end

L3 = input( 'Please enter the inductance for the third
harmonic filter (nH): ');

if (L3 <= 0)
L3=1e-9; %Third harmonic filter inductance
else
L3=L3 * 1e-9;
end

if harmonics == 2;
L5 = input( 'Please enter the inductance for the fifth
harmonic filter (nH): ');
if (L5 <= 0)
    L5=1e-9; %Fifth harminic filter inductance
else
L5=L5 * 1e-9;
end
else
L5 = 0 ;
end

if harmonics == 1
%third harmonic peaking circuit coefficients
dV=2;
gV=1.1547;
di=2.91;
gi=1.4142;
else
%5 resonators circuit coefficients
dV=2;
gV=1.2071;
di=3;
gi=1.5;
end

%Calculate optimum resistance and waveforms
RL=gV^2*Vdc^2/(2*Po);
Vom=gV*Vdc;
vdm=dV*Vdc;
Idc=gV*Vdc/(gi*RL);
idm=di*Idc;
A=2*acos(Idc/(Idc -idm));

%Calculate Filters
%base
```

```
C0=1/((2*pi*fo)^2*L0);
%3rd harmonic
C3=1/((2*pi*3*fo)^2*L3 );
if harmonics==2;
    C5=1/((2*pi*5*fO)^2*L3 );

end
%Display calculated values
fprintf ('\nRESULTS :\n');
fprintf ('RL % .2f Ohm \n', floor(100* RL + 0.5)/100);
fprintf ('L0 % .2f nH \n', floor(1e11 * L0 + 0.5)/100);
fprintf ('C0 % .3f pF \n', floor(1e14 * C0 + 0.5)/100);
fprintf ('L3 % .2f nH \n', floor(1e11 * L3 + 0.5)/100);
fprintf ('C3 % .3f pF \n', floor(1e14 * C3 + 0.5)/100);
if harmonics==2;
    fprintf ('L5; % .2f nH \n', floor(1e11 * L5 + 0.5)/100);
    fprintf ('C5; % .3f pF \n', floor(1e14 * C5 + 0.5)/100);

end

fprintf ('Idc % .2f mA \n', floor(1e5 *Idc+ 0.5)/100);
fprintf ('vdm % .2f v \n', floor(1e2 * vdm + 0.5)/100);
fprintf ('Vom % .2f V \n', floor(1e2 * Vom + 0.5)/100);
fprintf ('idm % .2f mA \n', floor(1e5 * idm + 0.5)/100);
```

6.5.1 CLASS S

Class S amplifier is based on switching of two transistors similar to Class D amplifier concept. The main difference between Class D and S amplifiers is in the way the amplifier is driven. According to the Class S amplifier operation, the analog input signal is converted into a digital pulse train via modulator instead of signal being alternately switched at the carrier frequency with a constant duty cycle. The fully digital pulse train then feeds the power-switching final-stage amplifiers, which in turn amplify it to the proper power level. In the ideal case, no overlapping occurs between current and voltage waveforms, and hence no power loss exists, which leads to a 100% efficiency independently of the power back-off. A demodulator is required at the output network to pick the required signal frequency and to restore the analog input signal. The amplifier can be implemented in VM or CM configurations. The VM configuration of Class S amplifiers with waveforms is given in Figure 6.23.

Example 6.3

Design Class E amplifier operating at optimum conditions for $f = 2.4\,\text{GHz}$, $Q_L = 7$, $P_o = 50$ [mW], and $V_{dc} = 1$ [V].

Solution: The following Matlab program is developed to design any type of switch-mode amplifier with GUI. The program script is given for Class E amplifier.
    Main Program: CLASS_E.m

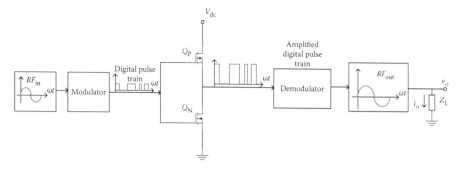

FIGURE 6.23 Voltage mode Class S amplifier configuration.

```
clc;
clear all ; %Clear all variables from workspace

prompt = {'Type "E" Class E,  "F" for Class F, "D" for Class D
and "DE" for Class DE"'};
dlg_title = 'Switch-mode Amplifier Design Program';
num_lines = 1;
def = {'1'};
answer = inputdlg(prompt,dlg_title,num_lines,def, 'on');

%convert the strings received from the GUI to numbers
valuearray=str2double(answer);

%Give variable names to the received numbers
choice=valuearray(1);

if (choice== 1)
CLASS_E; %Class-E PA design program
end

if (choice == 2)
classF ; %Class-F PA design program
end

if (choice == 3)
classd ; %Class-DE PA design program
end

if (choice == 4)
classde ; %Class-DE PA design program
end

function varargout = CLASS_E(varargin)

gui_Singleton = 1;
gui_State = struct('gui_Name',        mfilename, ...
```

```
                           'gui_Singleton',  gui_Singleton, ...
                           'gui_OpeningFcn', @CLASS_E_OpeningFcn, ...
                           'gui_OutputFcn',  @CLASS_E_OutputFcn, ...
                           'gui_LayoutFcn',  [] , ...
                           'gui_Callback',   []);
if nargin && ischar(varargin{1})
    gui_State.gui_Callback = str2func(varargin{1});
end

if nargout
    [varargout{1:nargout}] = gui_mainfcn(gui_State,
varargin{:});
else
    gui_mainfcn(gui_State, varargin{:});
end

% --- Executes just before CLASS_E is made visible.
function CLASS_E_OpeningFcn(hObject, eventdata, handles,
varargin)
% Choose default command line output for CLASS_E
handles.output = hObject;

% Update handles structure
guidata(hObject, handles);

% UIWAIT makes CLASS_E wait for user response (see UIRESUME)
% uiwait(handles.figure1);

% --- Outputs from this function are returned to the command
line.
function varargout = CLASS_E_OutputFcn(hObject, eventdata,
handles)

% Get default command line output from handles structure
varargout{1} = handles.output;

function input1_editText_Callback(hObject, eventdata, handles)
input = str2num(get(hObject,'String'));
if (isempty(input))
set(hObject,'String','0')
end
.
.
.

a1 = get(handles.input1_editText,'String');
a2 = get(handles.input2_editText,'String');
a3 = get(handles.input3_editText,'String');
a4 = get(handles.input4_editText,'String');
a5 = get(handles.input5_editText,'String');
a6 = get(handles.input6_editText,'String');
```

```
% length units in centimeters
% Length of core in centimeters
fo=str2num(a1);
QL=str2num(a2);
Po=str2num(a3);
Vdc=str2num(a4);
Vdssat=str2num(a5);
BVds=str2num(a6);


%Program begins

%Enter PA design parameters
%fo=input ('Please enter the frequency of operation, fo (MHz):
');
%if (fo<=0)
%fo=1e9; %Frequency
%else
%fo=fo*1e6;
%end

fo=fo*1e6;
w=2*pi*fo; %Radian frequency

if (QL < 1)
QL = 10; %Loaded quality factor
end


if (Po <= 0 )
Po=0.01; %Output power
else
Po=Po/1000;
end

if (BVds <= 0)
BVds = 0; %Breakdown voltage
end
%Calculate
Vsp =3.56 * Vdc ;

if (Vsp > BVds && BVds > 0)
    NVdc = BVds/3.56;
fprintf('Warning: Vdc should not exceed %.2f or the transistor
will go into breakdown!\n' , NVdc );
end
```

```
RL=0.577*(Vdc - Vdssat)^2/Po ; %Load resistance
L2=1e9*(QL*RL/w); %Series inductance
C1=1e12*(1/(w*RL*5.447)); %Parallel capacitance
C2=(C1* (5.447/QL)*(1+1.421/(QL - 2.08))); %Series Capacitance

Idc=1e3*(Vdc/(1.734*RL)); %DC current consumpt ion
isp=(2.86*Idc); %Maximum collector current dissipation


%DISPLAY THE RESULTS ON THE WINDOW
c1=num2str(RL);
c2=num2str(L2);
c3=num2str(C2);
c4=num2str(C1);
c5=num2str(Idc);
c6=num2str(Vsp);
c7=num2str(isp);

% need to convert the answer back into String type to display
it
set(handles.answer1_staticText,'String',c1);
set(handles.answer2_staticText,'String',c2);
set(handles.answer3_staticText,'String',c3);
set(handles.answer4_staticText,'String',c4);
set(handles.answer5_staticText,'String',c5);
set(handles.answer6_staticText,'String',c6);
set(handles.answer7_staticText,'String',c7);

guidata(hObject, handles);

% --- Executes during object creation, after setting all
properties.
function axes5_CreateFcn(hObject, eventdata, handles)
% hObject handle to axes5 (see GCBO)
% eventdata  reserved - to be defined in a future version of
MATLAB
% handles empty - handles not created until after all
CreateFcns called

% Hint: place code in OpeningFcn to populate axes5
axes(hObject)
imshow('classE.jpg')
```
When program runs, the output is

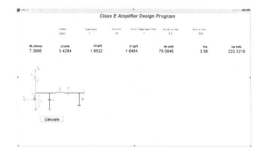

Design Example: Class E Amplifier

Design single-ended Class E amplifier at 13.56 MHz with 5% bandwidth for semiconductor wafer processing applications. The amplifier should be able to provide 150 W at the center frequency. The spurious level of the output signal should be kept at –30 dB for all load conditions. The transistor that will be used is IXFH12N100F or any other transistor with similar characteristics. The complete design specifications are given Table 6.5.

Solution: To meet with the desired RF amplifier driver requirements, the topology shown in Figure 6.24 is proposed. This topology differs from the existing topology by incorporating the in-line and off-line filters to eliminate spurious oscillations and harmonic levels.

TABLE 6.5
Design Specifications for Class E Amplifier

Load Condition	Minpower [W]	Harmonics [dBC]	Spurious [dBC]	Rail Voltage [B+]	Frequency [MHz]
1:1	150	–30	–30	0–120 V	13.56 ± 5%
3:1	60	–30	–30	0–120 V	13.56 ± 5%
5:1	30	–30	–30	0–120 V	13.56 ± 5%
Inf:1	30	–30	–30	0–120 V	13.56 ± 5%

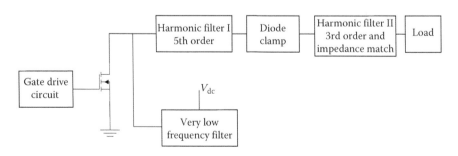

FIGURE 6.24 Class E amplifier with inductive clamp circuit topology is extended to include effective filtering.

The circuit in Figure 6.24 has an inductive clamp circuit [24] and effective filtering for oscillations and harmonics. The details of design for the subsections shown in Figure 6.24 are given in the following section.

Off-line Very Low-Frequency Filter: Network I
This filter is connected at the drain of the MOSFET with a 25-Ω load termination. It presents the MOSFET high impedance with a minimum resistance at the operational frequency and presents approximately 25 Ω of resistance with minimum reactance from 100 kHz to 5.4 MHz. The filter is a fifth-order Chebyshev filter with 0.01 dB passband ripple. Its 3 dB point is chosen to be 7.1 MHz. The first element of the filter is a series inductor, which is an appropriate component for the topology we use in this amplifier. If the dual of this filter is used, the shunt capacitor would be the first element. As a result, the transistor would see a capacitive loading at the operational frequency, adding to the output capacitance C_{oss} of the transistor. The additional capacitance contributing to output capacitance of the transistor would be approximately 900 pF. The impedance and corresponding reactive component values that this filter presents to the MOSFET at the operational frequencies are given in Table 6.6.

In-line Harmonic Rejection Filter with Clamping Diode: Network II
The design details for the in-line harmonic filter constitute the main body of the resonant network with the shunt inductance presented by Network I. The in-line harmonic filter has to be adjusted to include the inductive clamping diode network. One way to do that is to make the inductive clamping diode part of the harmonic filter. The harmonic filter is a fifth-order Chebychev filter with 0.01 dB passband ripple. The harmonic attenuation is enhanced, and proper impedance matching is accomplished by addition of a three-pole PI network between the overall network and the load. This network is shown in Figure 6.25.

The impedance that Network II presents the MOSFET is found to be inductive for all operational frequencies as expected. This guarantees that whatever the load condition is, the transistor always sees an inductive loading. This indicates

TABLE 6.6

Impedances and Inductance Values at f_{op}

Frequency [MHz]	Impedance [Ω]	Inductance [nH]
5.4	25+1.65i	48.63
12.882	2.5E–002+36.2i	447.24
13.56	2.5E–002+39.65i	465.38
14.238	2.5E–002+42.025i	469.76

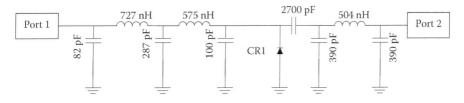

FIGURE 6.25 In-line harmonic rejection filter with inductive clamping diode—Network II.

TABLE 6.7

Impedances and Inductances at f_{op}

Frequency [MHz]	Impedance [Ω]	Inductance [nH]
16.75	75.75 + 0.6i	5.70
12.882	23.97 + 2.025i	25.02
13.56	21.1 + 6.325i	74.24
14.238	20.97 + 12.05i	134.70

that the diode network is functioning properly. The impedance and corresponding reactive component values at the operational frequencies are given in Table 6.7.

Final Design

The final design interfaces Network I, Network II, and the active device MOSFET. The finalized network is simulated by a nonlinear circuit simulator (Pspice). Although we need to tune some of the component values to obtain better performance, the calculated component values are found to be close to the final design component values. The close approximation between calculated and final component values requires the knowledge of the accurate C–V characteristics of the transistor used. If C–V characteristics of the transistors are not known, the final component values in the design may deviate from the calculated component values significantly. The final circuit is shown in Figure 6.26.

The zero voltage switching (ZVS) is accomplished as it is seen from the waveforms for V_{ds} and V_{load} shown in Figure 6.27. The simulated drain efficiency is found to be between 90% and 92% at the operational frequencies within the bandwidth. One of the most important features of the inductive clamp topology is the protection that it provides to the transistor. In this way, the amplifier protects itself to any transient conditions that can cause high dissipation in the die and result in destruction. The functionality of the topology is verified with the simulation for different load conditions at the rated DC supply voltage, $V_{dc} = 120$ V. The die dissipation with inductive clamping diode is simulated and shown in Figures 6.28.

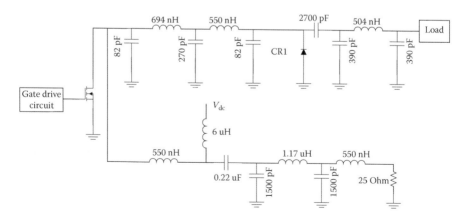

FIGURE 6.26   The final circuit that is simulated and built.

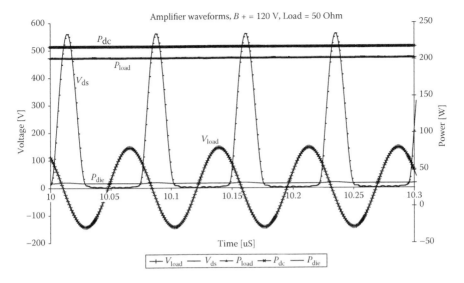

FIGURE 6.27   Simulated amplifier waveforms when load is 50 Ω.

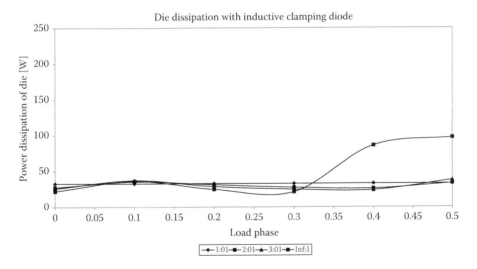

FIGURE 6.28   Simulated die dissipation with inductive clamp.

When the amplifier is integrated with inductive clamp circuit, the die dissipation is extremely low, and as a result, the destruction of die to any load condition is prevented. The simulated waveforms including drain-to-source and output signals for 50 Ω are given in Figure 6.29.

Experimental Results
The amplifier shown in Figure 6.26 was built, and the results were compared with the simulated results and verified that they comply with design specifications given in Table 6.5. The first step was to see whether the RF amplifier driver

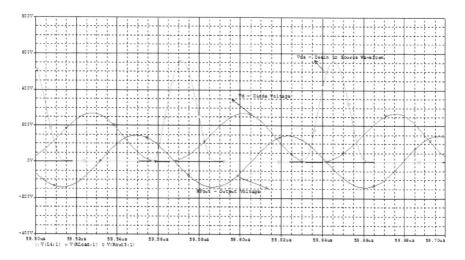

FIGURE 6.29   Simulated results for V_{ds} and V_{out}.

complies with design specifications. Table 6.8 gives the measured results for spurious and harmonic levels for operational frequencies within the bandwidth when the load is 50 Ω.

The worst spurious case was measured at 13.56 MHz, for loads with VSWR = 3 : 1. The measured spurious levels for three frequencies within the operation bandwidth are plotted in Figure 6.30. The spur levels even for the worst condition are found to be within the specification. The measured drain efficiency is given in Figure 6.31. The drain efficiency ranges from 87% to 88%. The die dissipation is found to be very low for all frequencies, and it is consistent with simulation results.

Another important power amplifier performance parameter is the delivered power distribution over frequency. This performance parameter is shown in Figure 6.32.

When the amplifier is designed, it should be made sure that the die dissipation and as a result the heat generated by the transistors are controllable with the system cooling that is in place. The thermal cooling for high-power amplifiers mostly consists of both convection- and conduction-type cooling system. The die dissipation contours for this amplifier for various load conditions are tested and given in Figure 6.33. The color map shows the amount of dissipation on the Smith chart. The measured waveforms are given in Figure 6.34. The blue color shows the waveform taken at diode, and yellow shows the drain-to-source voltage measurements on the load-line for an open load condition.

Design Example 2: Push–Pull Class E Amplifier

Design Class E amplifier in push–pull configuration to provide minimum P_{out} = 2500 W at 180 V using three transistors on each of the push and pull sides at 13.56 MHz. Connect input and output via balun with 25 Ω impedance. The transistor presents about 12.5 Ω at the operational frequency with rated voltage. Use IXFH12N100F transistors that are rated for 1200 V. Input power should not exceed 150 [W].

TABLE 6.8

Measured Results for Spurious and Harmonic Levels

		Spurs		Harmonic Levels	
$F = 13.56\,\text{MHz}$	V_{dc} [V]	dBC	Spur Frequency	2nd Harmonic	3rd Harmonic
	25	−53.57	Noise floor	−34.71	−44.47
	50	−52.84	Noise floor	−36.54	−47.56
	75	−60.78	Noise floor	−37.08	−48.08
	100	−61.14	Noise floor	−36.93	−47.72
	120	−60.07	Noise floor	−37.45	−46.17
$F = 12.882\,\text{MHz}$	V_{dc} [V]	dBC	Frequency	2nd Harmonic	3rd Harmonic
	25	−66.9	Noise floor	−48.03	−57.58
	50	−64.94	Noise floor	−35.76	−45.77
	75	−58.25	Noise floor	−36.72	−47.1
	100	−59.24	Noise floor	−37.020	−46.96
	120	−59.24	Noise floor	−36.78	−46.1
$F = 14.238\,\text{MHz}$	V_{dc} [V]	dBC	Frequency	2nd Harmonic	3rd Harmonic
	25	−66.01	Noise floor	−37.02	−46.38
	50	−59.34	Noise floor	−38.56	−51.23
	75	−62.94	Noise floor	−39.13	−51.79
	100	−65.22	Noise floor	−39.08	−52.43
	120	−67.27	Noise floor	−39.12	−51.82

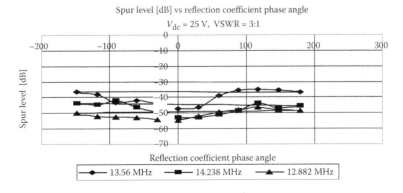

FIGURE 6.30 Measured spurious levels for the worst condition.

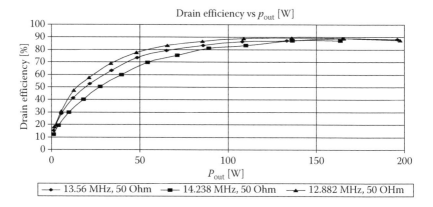

FIGURE 6.31   Measured drain efficiency for three frequencies.

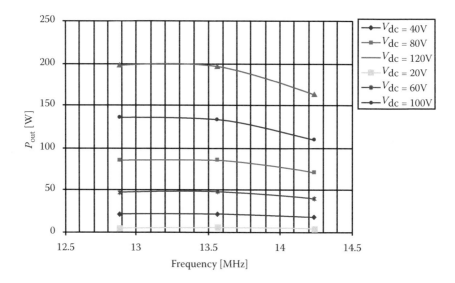

FIGURE 6.32   Delivered power distribution over frequency.

Solution: The Class E amplifier circuit is designed and simulated as shown in Figure 6.35.

The simulated results for the waveforms are given in Figures 6.36 through 6.38. Figure 6.36 shows V_{ds}, and Figure 6.37 gives the output current. Figure 6.38 illustrates the output voltage.

The simulated values show that the output power for 50 Ω is equal to 2745 [W]. The impedance has been simulated at the points shown on the amplifier circuit shown in Figure 6.35. As seen from Table 6.9, the impedances at the drain of the transistor are very close to the desired impedance value at 13.56 MHz.

The design can now be implemented. The design of input balun is given for reference design. In practice, each transistor needs to be fed with an isolation gate transformer. There is a reference design for isolation gate transformer for the push–pull amplifier shown in Figure 6.35.

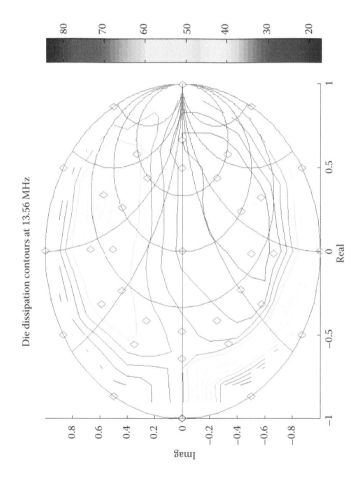

FIGURE 6.33 Die dissipation contours for Class E amplifiers.

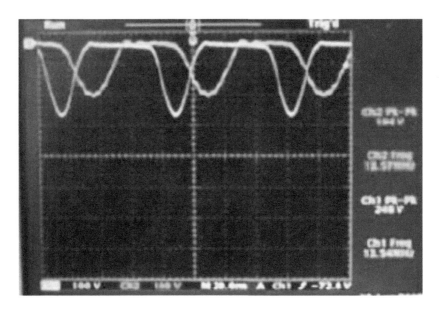

FIGURE 6.34 Measured waveforms for V_{ds} (yellow) and diode (blue) for Class E amplifiers for open load.

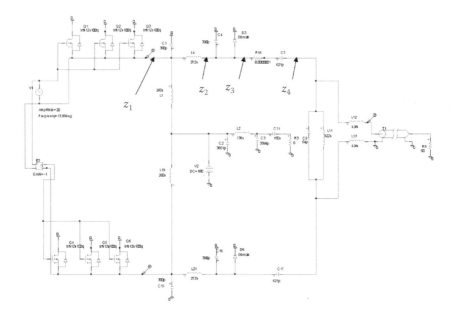

FIGURE 6.35 Push–pull Class E amplifier designed at 13.56 MHz.

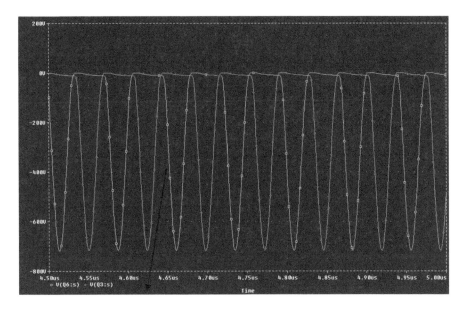

FIGURE 6.36 V_{ds} waveform for push–pull Class E amplifier.

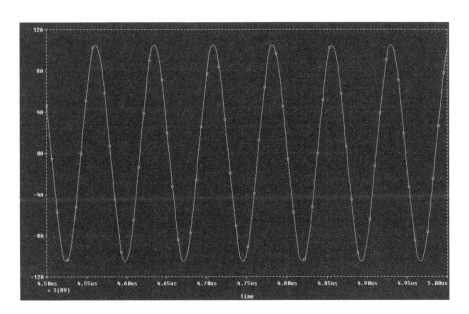

FIGURE 6.37   Output current waveform for push–pull Class E amplifier.

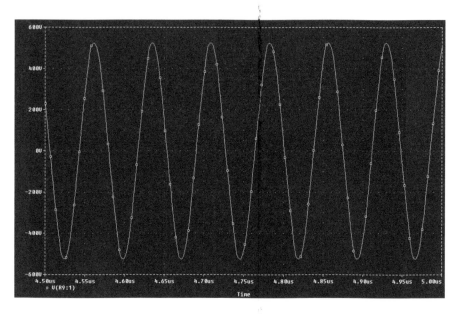

FIGURE 6.38   Output voltage waveform for push–pull Class E amplifier.

TABLE 6.9
Simulated Node Impedances

Impedance Node	$V < \Theta$	$I < \Theta$	$Z < \Theta$
Z_1	$298.9 < -55.37°$	$23.71 < -117.2°$	$12.60 < 61.83°$
Z_2	$265.1 < -156.6°$	$18.81 < -108.9°$	$14.09 < -47.7°$
Z_3	$264.7 < -156.6°$	$11.14 < -131.3°$	$23.76 < -25.3°$
Z_4	$257.9 < -109.1$	$10.33 < -109.1$	$24.97 < 0°$

Design of Input Balun
Based on the given information, the equivalent circuit of the input balun can be illustrated as shown in Figure 6.39.

For maximum power transfer, $R_L = R_s$. It is given that $0.5R_L = 25$. Then,

$$R_L = R_s = 50 \ [\Omega]$$

It is also given that $P_{in} = 150\,W$. This requires the operational rms voltage and current values at the input to be

$$P_{in} = 150 \ W, \quad Z_{in} = 50 \ \Omega, \quad V_{in,rms} = 86.6 \ V \text{ and } I_{in,rms} = 1.73$$

At the output of the balun, we will have

$$P_{out,1} = 75 \ W, \quad Z_{out,1} = 25 \ \Omega, \quad V_{out,rms,1} = 43.3 \ V, \quad I_{out,rms,1} = 1.73 \ A$$

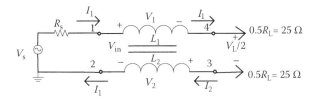

FIGURE 6.39 The equivalent circuit of input balun.

The magnetic core that will be used in the design of balun will be selected based on the impedance and flux density requirement. The rule of thumb in the design of balun is to have impedance value at 5–10 times higher than the highest impedance termination at any of its ports. The highest impedance is at the input port of the balun and is equal to 50 W. The target impedance value for the balun is

$$Z_{req} = 2\pi f L = 2\pi(13.56 \times 10^6) \times L = 500 \ \Omega$$

Then,

$$L \geq \frac{500}{2\pi(13.56 \times 10^6)} \quad \text{or} \quad L \geq 5.868 \ \mu H.$$

The inductance is found as

$$L = \frac{4\pi N^2 \mu_i A_{TC}}{l_e} [nH]$$

We can now find the number of turns, N, if we have the characteristics of the core that we will be using. We determine to use core material manufactured by Ferronics with part number 11-260. This is K-type material with permeability of 125, $\mu_i = 124$, OD = 1.27 cm, ID = 0.714 cm, and $h = 0.478$ cm. The saturation flux density at 10 MHz is given as 3200 [G]. This value is expected to significantly drop as frequency goes up. We will be using 10 cores by stacking 5 on each side as shown in Figure 6.40.

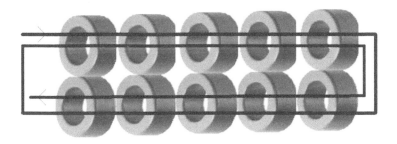

FIGURE 6.40 Construction of the balun using multiple cores.

The number of turns is found as

$$N = \sqrt{\frac{L \times l_e}{4\pi\mu_i A_{Tc}}} = \sqrt{\frac{5868 \times 3.03}{4\pi \times 125 \times 10 \times 0.132}} = 2.92 \approx 3 \text{ turns}$$

The operational flux density B_{op} can be found as

$$B_{op} = \frac{V_{rms} \times 10^8}{4.44 f N A_{Tc}} = \frac{86.6 \times 10^8}{4.44 \times 13.56 \times 10^6 \times 2.92 \times 10 \times 0.132} = 37.07 \text{ [Gauss]}$$

The operational flux density is well below the saturation flux density. It is also worth to note that increasing the number of the cores reduces the operational flux density. RG-188 coaxial cable is selected for winding since it is capable of carrying the current required and has $50\,\Omega$ characteristic impedance.

The final configuration of the balun that is constructed is shown in Figure 6.41. Matlab GUI program illustrating the calculated values for the design of a balun is given in Figure 6.42.

The final configuration illustrated in Figure 6.41 is constructed with three turns. When $N = 3$, using the inductance formulation, inductance is found to be 6.16 [μH]. The inductance of balun is measured using HP 4191A, RF impedance analyzer. The setup to measure the inductance of the balun is shown in Figure 6.43.

The self-inductance value is measured to be 6.4 [μH] and close to the calculated value.

The frequency response of the input balun is measured with HP 8753B network analyzer. The measurement setup and measured response are illustrated in Figures 6.44 and 6.45, respectively. The characteristics of some of the coaxial cables that can be used in the construction of balun are given in Table 6.10.

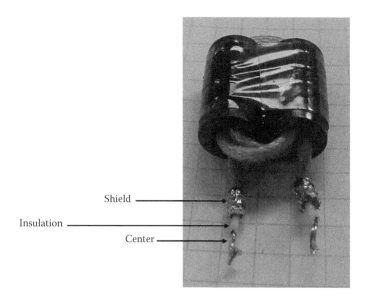

Shield

Insulation

Center

FIGURE 6.41   Balun that is constructed.

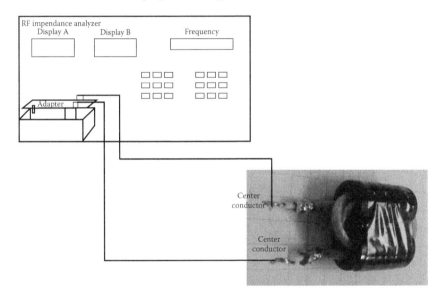

FIGURE 6.42 Matlab GUI program to design Balun.

FIGURE 6.43 Measurement setup for inductance measurement of input balun input.

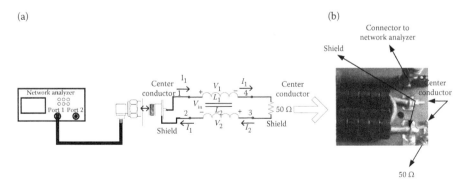

FIGURE 6.44 (a) Block diagram for measurement of the frequency response of input balun and (b) Measurement setup for frequency response for input balun.

TABLE 6.10
Coax Cable Information

Coax Cable	OD (in.)	Impedance	Jacket Material	Center Conductor Material	Center Conductor Type	Center Conductor Diameter (in.)	Dielectric
RG142	0.195	50	Foamed polyethylene	Silver plated copper steel wire	Solid	0.037	PTFE
RG178	0.071	50	Foamed polyethylene	Silver plated copper steel wire	Stranded	0.012	PTFE
RG179	0.1	75	Foamed polyethylene	Silver plated copper steel wire	Stranded	0.012	PTFE
RG188	0.1	50	PTFE	Silver plated copper steel wire	Stranded	0.0201	PTFE
RG316	0.098	50	Foamed polyethylene	Silver plated copper steel wire	Stranded	0.0201	PTFE
RD316	0.114	50	Foamed polyethylene	Silver plated copper steel wire	Stranded	0.0201	PTFE
RG400	0.2	50	Foamed polyethylene	Silver plated copper steel wire	Stranded	0.038	PTFE
RG393	0.39	50	Foamed polyethylene	Silver plated copper steel wire	Stranded	0.094	PTFE
RG58	0.195	50	PVC-NC	Tinned copper	Stranded	0.036	PE
RG59	0.242	75	PVC-NC	Bare copper	Solid	0.023	PE
RG214	0.425	50	PVC-NC	Silver plated copper	Stranded	0.089	PE
RG393	0.212	50	PVC-NC	Silver plated copper	Solid	0.036	PE

PTFE, polytetrafluoroethylene

The measured values showing the frequency response of the input balun between 1 MHz and 30 MHz are given in the following table.

Frequency [MHz]	Real	Imaginary	Inductance [nH]
1	51.252	0.238	37.923
5	51.393	0.8789	27.976
13.56	51.734	2.1484	25.216
27.12	52.777	4.168	24.44

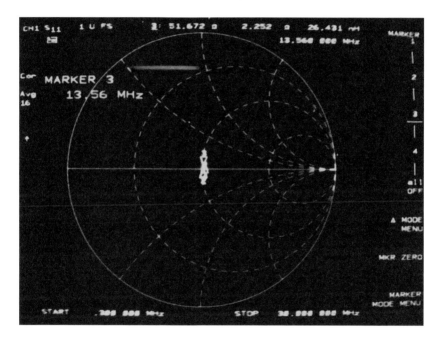

FIGURE 6.45 Frequency response of input balun.

Isolation Gate Transformer Design
It is required to design and characterize an isolation gate transformer to operate at 13.56 MHz. Current at the primary is given to be 1.73 Arms. The power input is 150 [W], and the voltage at the secondary is desired to be 30 [Vrms]. The input impedance that will be presented is 50 Ω

Based on the given information, the voltage at the secondary is 30 V, and at the primary is

$$V_p = \sqrt{150(50)} = 86.6 \text{ [Vrms]}$$

The turn ratio is then calculated to be,

$$\frac{N_1}{N_2} = \frac{V_p}{V_s} = \frac{86.6}{30} = 2.89$$

This can be rounded to 3. Hence, the transformer with ratio 3:1 should satisfy the design specifications. Consider single-bead geometry given in Figure 6.46. The cross-sectional area and magnetic path length are given as

$$A_e = 0.0996 \text{ cm}^2, \, l_e = 1.5 \text{ cm}$$

The permeability of the material is 125. In our design, we did not initially know the number of turns. So, as a rule of thumb, we targeted impedance on the primary 10 times higher than the impedance that is interfaced.

$$Z = 2\pi f L = 2\pi(13.56 \times 10^6) \times L = 500 \, \Omega$$

From this relation, the target inductance is 5.86 µH. We use the following formulation to obtain the inductance. When six beads are used, the total cross-sectional area becomes $A_{Te} = 6 \times A_e$, and the inductance value becomes 5.63 µH.

$$L = \frac{0.4\pi\mu_i A_{Te}(\text{cm}^2)N^2}{100 l_e(\text{cm})} [\mu H]$$

When this is placed back into impedance formulation, we obtain 3:1 transformer with six beads is realized using the following bead configuration shown in Figure 6.47.

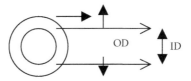

FIGURE 6.46 Single-bead geometry.

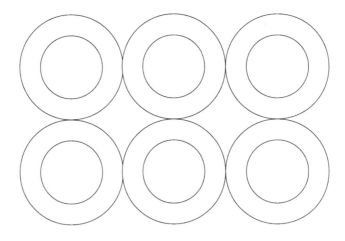

FIGURE 6.47 Six-bead geometry.

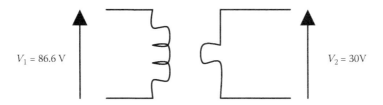

FIGURE 6.48   Equivalent circuit six-bead geometry.

FIGURE 6.49   Constructed isolation gate transformer.

$$Z = 2\pi f L = 2\pi(13.56 \times 10^6) \times L = 479.8 \ \Omega$$

The size of the wires can be determined as follows. Because the current at the primary is 1.73A, the min wire size should be 26AWG. 26AWG wire will be able to handle 2.2A if it is solid wire. Because the current at the secondary is 5.19A, the min wire size should be 22AWG. 22AWG will be able to handle 5.5A when solid wire is used. One of the design practice is to choose the wire insulation properly to prevent arcing between primary and secondary windings.

The equivalent circuit for the isolation transformer is given in Figure 6.48. The voltage difference between the windings is $\Delta V = 86.6 - 30 = 56.6$ V. Hence the breakdown voltage of the wires is greater than 56.6V. The Teflon insulation used on the primary winding and enameled wire on the secondary winding has much more higher voltage rating than 56.6V. The transformer is implemented in Figure 6.49. 14AWG wire is used on the secondary side, which has much more higher rating than desired value. This transformer is used to drive the MOSFETs and is connected between gate and source of the transistor.

Characterization of the transformer is done using open and short circuit tests as explained earlier. The open and short circuit test results for this transformer is shown in Table 6.11.

The measured values are

$$L_p = 5.15[\mu H], \ L_s = 579.6[nH], \ L'_s = 22[nH], \ n = 3$$

Using the equation,

$$\left(\frac{L_{1l}^2}{n^2 L_p}\right) - \frac{2L_{1l}}{n^2} + \left(L'_s - L_s + \frac{L_p}{n^2}\right) = 0$$

or

$$\left(\frac{L_{1l}^2}{9(5.15 \times 10^{-6})}\right) - \frac{2L_{1l}}{9} + \left(22 \times 10^{-9} - 579.6 \times 10^{-9} + \frac{5.15 \times 10^{-6}}{9}\right) = 0$$

TABLE 6.11

Open and Short Circuit Test Results for Isolation Gate Transformer

Measured	Primary	Secondary	L
Primary	Open	Open	$5.15\,\mu H$
Secondary	Open	Open	$579.6\,nH$
Secondary	Short	Open	$22\,nH$

The solution of the foregoing equation gives the leakage inductance on the primary side as $L_{1l} = 66.22\,[\text{nH}]$. The leakage inductance on the secondary side can be found from

$$L_s = L_{2l} + \frac{L_p - L_{1l}}{n^2}$$

The leakage inductance on the secondary is then equal to

$$L_{2l} = L_s - \frac{L_p - L_{1l}}{n^2} = 579.6\,[\text{nH}] - \frac{5150\,[\text{nH}] - 66.22\,[\text{nH}]}{9} = 14.74\,[\text{nH}]$$

Hence, the measured leakage inductances for the isolation gate transformer are

$$L_{1l} = 66.22\,[\text{nH}] \text{ and } L_{2l} = 14.74\,[\text{nH}]$$

The self-inductances on the primary and secondary sides are found from

$$L_{1p} = L_p - L_{1l} = 5150\,[\text{nH}] - 66.22\,[\text{nH}] = 5083.78\,[\text{nH}]$$

$$L_{2s} = L_s - L_{2l} = 579.6\,[\text{nH}] - 14.74\,[\text{nH}] = 564.86\,[\text{nH}]$$

The coupling coefficient is obtained from

$$k = \sqrt{\frac{L_{1p}L_{2s}}{L_p L_s}} = \sqrt{\frac{(5083.78)(564.86)}{(5150)(579.6)}} = 0.98$$

which is very close to 1 as desired. The mutual inductance is then equal to

$$M = k\sqrt{L_p L_s} = 0.98\sqrt{(5150)(579.6)} = 1693.14\,[\text{nH}]$$

REFERENCES

1. G. Giustolisi, A. D. Grasso, S. Pennisi, High-drive and linear CMOS Class-AB pseudo-differential amplifier, *IEEE Transactions on Circuits and Systems II: Express Briefs*, Vol. 54, No. 2, pp.112–116, 2007.

2. H. Kosugi, T. Matsumoto, T. Uwano, A high-efficiency linear power amplifier using an envelope feedback method, *Electronics and Communications in Japan, Part 2*, Vol. 77, No. 3, pp. 50–57, 1994.

3. G. Vendelin, A. Pavio, U. Rohde, Microwave Circuit Design using Linear and Nonlinear Techniques, 2nd ed. John Wiley & Sons, Hoboken, NJ, 2005.

4. M. Albulet, *RF Power Amplifiers*. Noble, Atlanta, GA, 2001.

5. S. Goto, T. Kunii, T. Oue, K. Izawa, A. Inoue, M. Kohno, T. Oku, T. Ishikawa, A low distortion 25 W Class-F power amplifier using internally harmonic tuned FET architecture for 3.5 GHz OFDM applications, *Proceedings of IEEE MTT–S International Microwave Symposium Digest, 2006*, pp. 1538–1541, June 2006.

6. F.-Y. Chen, J.-F. Chen, R.-L. Lin, Low-harmonic push- pull Class-E power amplifier with a pair of LC resonant networks, *IEEE Transactions on Circuits and Systems I: Regular Papers*, Vol. 54, No. 3, pp. 579–589, 2007.

7. F. J. Ortega-Gonzlez, Load-pull wideband Class-E amplifier, *IEEE Microwave and Wireless Components Letters*, Vol. 17, No. 3, pp. 235–237, 2007.

8. W. Saito, T. Domon, I. Omura, M. Kuraguchi, Y. Takada, K. Tsuda, M, Yamaguchi, Demonstration of 13.56-MHz Class-E amplifier using a high- voltage GaN power-HEMT, *IEEE Electron Device Letters*, Vol. 27, No. 5, pp. 326–328, 2006.

9. H.-S. Oh, T. Song, E. Yoon, C.-K. Kim, A powerefficient injection-locked class-E power amplifier for wireless sensor network, *IEEE Microwave and Wireless Components Letters*, Vol. 16, No. 4, pp. 173–175, 2006.

10. S. C. Cripps, *RF Power Amplifiers for Wireless Communication*, Artech House, Inc., Norwood, MA, 1999.

11. H. L. Krauss, C. W. Bostian, F. H. Raab, *Solid State Radio Engineering*, John Wiley & Sons, New York, 1980.

12. F. H. Raab, P. Asbeck, S. Cripps, P. B. Kenington, Z. B. Popovic, N. Pothecary, J. F. Sevic, N.O. Sokal, Power amplifiers and transmitters for RF and microwave, *IEEE Transactions on Microwave Theory Techniques*, Vol. 50, No. 3, pp. 814–826, 2002.

13. N. O. Sokal, A. D. Sokal, Class EA new class of high-efficiency tuned single-ended switching power amplifiers, *IEEE Journal of Solid-State Circuits*, Vol. 10, No. 3, pp. 168–176, 1975.

14. S. Sivakumar, A. Eroglu, Analysis of Class-E based RF power amplifiers using harmonic modeling, *IEEE Transactions on Circuits and Systems I : Regular Papers*, Vol. 57, No. 1, pp. 299–311, 2010.

15. F. Raab, Idealized operation of the Class-E tuned amplifier, *IEEE Transactions on Circuits and Systems*, Vol. CAS-24, No. 2, pp. 239–247, 1978.

16. N. O. Sokal, Class-E RF power amplifiers, QEX, pp. 9–20, January/February 2001.

17. M. Kazimierczuk, K. Puczko, Exact analysis of Class-E amplifier at any Q and switch duty cycle, *IEEE Transactions of Circuits and Systems*, Vol. 34, pp. 149–159, 1987.

18. M. K. Kazimierczuk, K. Puczko, Class-E tuned power amplifier with antiparallel diode or series diode at switch, with any loaded Q and switch duty cycle, *IEEE Transactions on Circuits and Systems*, Vol. 36, No. 9, pp. 1201–1209, 1989.

19. H. Sekiya, I. Sasase, S. Mori, Computation of design values of Class E amplifiers without using waveform equations, *IEEE Transactions on Circuits and Systems I: Fundamental Theory and Applications*, Vol. 49, No. 1, pp. 966–978, 2002.

20. M. Acar, A. J. Annema, B. Nauta, Analytical design equations for Class-E power amplifiers, *IEEE Transactions on Circuits and Systems-I: Regular Papers*, Vol. 54, No. 12, pp. 2706–2717, 2007.

21. Y. B Choi, K. K. M. Cheng, Generalised frequency-domain analysis of microwave Class-E power amplifiers, *IEE Proceedings Micorwave Antennaas Propagation*, Vol. 148, No. 6, pp. 403–409, 2001.

22. IRF450 Transistor Datasheet. http://www.irf.com/product-info/datasheets/data/jantx2n6770.pdf.
23. F. Raab, Maximum efficiency and output of Class-F power amplifiers, *IEEE Transactions on Microwave Theory Technology*, Vol. 49, No. 6, pp. 1162–1666, 2001.
24. A. Eroglu, A. Radomski, D. Lincoln, Y. Chawla, Improvement of Class E amplifier with inductive clamp circuit topology and its applications, 2006 IEEE MTT-S, pp. 906–909, June 2006.

7 Phase-Controlled Switch-Mode Amplifiers

7.1 INTRODUCTION

In this chapter, the harmonic modeling of Class E amplifiers discussed in Chapter 6 is extended to phase-controlled Class E power amplifier (PA) pairs to study the impact of phase control on the voltage, current, and power control profiles. Phase control of amplifier pairs enhances their controllability and adds flexibility to their power control characteristics [1–7]. However, the phase control operation results in asymmetric operation of the individual amplifiers of the pair [4]. Hence, the impact of the asymmetric operation, in light of the nonlinear characteristics of the Class E amplifier, on the component stress profile needs to be carefully studied while considering the use of phase-controlled Class E amplifier pairs as potential solution to specific applications. A detailed analysis of the phase-controlled Class E amplifier using circuit waveform equations for standard load conditions, assuming sinusoidal resonant tank current, is dealt with in Reference [2]. Reference [4] extends the prior work of Reference [8] to compute the design values for the phase-controlled Class E amplifier pair, without using explicit waveform equations as discussed and detailed in Reference [9]. A load-pull analysis of the phase-controlled Class E amplifier pair is discussed in Reference [7], which is based on the analysis of the independent amplifiers, with the equilibrium for the amplifier pair achieved at load points at which the load-pull seen by one amplifier is equal to that seen by the other.

7.2 PHASE CONTROL OF A CLASS E POWER AMPLIFIER PAIR

Figure 7.1 shows the circuit topology of the phase-controlled Class E PA pair, where two single-ended Class E PAs are combined through a transformer-combiner circuit to provide the total radio frequency (RF) output power. Each PA is rated to provide half the total rated power. The power switches of the two PAs are controlled in synchronism with two independent RF gate drive pulse signals of fixed and identical duty cycles (nominally 50%), however, with a facility to vary the phase of the pulse signals with reference to each other. The variation of the phase between the two gate drive signals facilitates the RF output of one of the PAs to be phase-shifted from the other. As a result, RF output through the transformer-combiner circuit is the combined total of two phase-shifted RF powers, one from each PA. By controllably varying the phase difference between the two drive pulse signals from 0° through 180° using a phase control signal, the total RF output power can be varied from zero through maximum, with the maximum power determined by the DC input rail voltage, V_{dc}, and the load impedance.

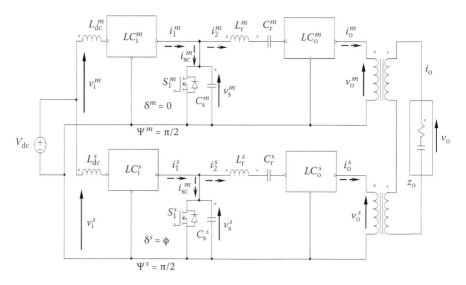

FIGURE 7.1   Circuit configuration of the phase-controlled Class E power amplifier pair.

The equations for the phase-controlled Class E power amplifier pair (Figure 7.1) can be written by combining two sets of equations—one for the first PA, the master (denoting the corresponding parameters and variables by a superscript m), and the other for the second, the slave (denoting the corresponding parameters and variables by a superscript s). The phase angle of the master PA, δ^m, can be chosen as zero, and that of the slave, δ^s, can be chosen a desired value toward meeting RF output power-control objective.

Assuming that the combiner transformers shown in Figure 7.1 are ideal and have a 1:1 transformation ratio, the equation relating the common output current i_o and the output voltages v_o^m and v_o^s of the master and slave PAs, respectively, is given by

$$\begin{pmatrix} v_{odh} \\ v_{oqh} \end{pmatrix} = \begin{pmatrix} v_{odh}^m + v_{odh}^s \\ v_{odh}^m + v_{odh}^s \end{pmatrix} = \begin{pmatrix} z_{cdh} & -z_{cqh} \\ z_{cqh} & z_{cdh} \end{pmatrix} \begin{pmatrix} i_{odh} \\ i_{oqh} \end{pmatrix} \tag{7.1}$$

An exact solution for the phase-controlled Class E PA pair configuration is obtained by solving a system of resulting simultaneous nonlinear equations with v_{idh}^m, v_{iqh}^m, v_{idh}^s, and v_{iah}^s all set to zero, and both v_{s0}^m and v_{s0}^s set to V_{dc}.

Because the simultaneous equations derived earlier for the Class E PA configurations are nonlinear, the iterative process for their solution is considered. It can be expressed that

$$\sum_{k=1}^m g_k(\varepsilon_0, \varepsilon_1, \varepsilon_2)u_k \tag{7.2}$$

where $g_k()$ is a nonlinear function of the angles ε_i, $i = 0, 1, 2$, and u_k, $k = 1, 2, ..., m$ are the circuit variables made up of the input DC current, the direct and quadrature axes components of the circuit voltages and currents, and their harmonics. Note that for fixed values of the angles ε_i, $i = 0, 1, 2$, these equations have a linear form. Hence, the following iterative process is considered: Set the initial estimates ε_i, $i = 0, 1, 2$ to zero; solve the set of resulting linear equations using matrix inversion to compute the direct and quadrature axes harmonic components of the various nodal voltages and branch currents; reconstruct the switch voltage waveform from the solution obtained in the previous step, and compute the actual ε_i, $i = 0, 1, 2$, and compare with their corresponding presolution estimates; update the estimates using an appropriate criterion; and repeat the process until convergence, using equality, within an acceptable tolerance of the pre- and postsolution values of ε_i, $i = 0, 1, 2$ within an iteration, as the convergence criterion.

For the case of single-ended Class E amplifier, for a given ε_i, $i = 0, 1, 2$, equations can be combined into the following matrix forms:

$$\bar{A}\bar{x} = \bar{b} \tag{7.3}$$

where $\bar{x} = \left[i_{dc} \ (i_1)_{nh} \ (v_s)_{nh} \ (v_o)_{nh} \ (i_o)_{nh} \right]^T$ is the vector of independent variables of interest, where the form $(p)_{nh}$ represents the row vector $[p_{d1} \ p_{q1} \ p_{d2} \ p_{q2}, ..., p_{dn} \ p_{qn}]$, where p_{dh} and p_{qh}, respectively, represent the hth harmonic direct and quadrature axes components of the arbitrary circuit variable p; System matrix $\bar{A}$ is of the form

$$\bar{A} = \begin{bmatrix} \bar{A}_s(\varepsilon) \\ \bar{A}_{is} \\ \bar{A}_{so} \\ \bar{A}_o \end{bmatrix} \tag{7.4}$$

where $\bar{A}_s(\varepsilon)$ is a $(nh+1) \times (4nh+1)$ submatrix of coefficients that describes the switch voltage a given ε_i, $i = 0, 1, 2$; $\bar{A}_{is}$ is a $(nh) \times (4nh+1)$ submatrix of coefficients that describes the input-switch circuit interface; $\bar{A}_{so}$ is a $(nh) \times (4nh+1)$ submatrix of coefficients that describes the switch-output circuit interface; and $\bar{A}_o$ is a $(nh) \times (4nh+1)$ submatrix of coefficients that describes the output circuit. Note that the submatrices other than $\bar{A}_s(\varepsilon)$ have fixed elements. Vector $\bar{b}$ is of the form $\bar{b} = \left[V_{dc} \ (0)_{4nh} \right]^T$, where $(0)_{4nh}$ is a row vector of $4nh$ zero elements.

A similar matrix equation can be constructed for the phase-controlled Class E pair, with the following setting: $\psi^m = \psi^s = \pi/2$; $\delta^m = 0$; $\delta^s = \phi$, the desired or controlled phase angle. The resulting equation is of the form

$$\bar{A}^{ms}\bar{x}^{ms} = \bar{b}^{ms} \tag{7.5}$$

where

$$
\overline{A}^{ms} =
\begin{bmatrix}
\overline{A}_{11}^{m} & 0 & \overline{A}_{12}^{m} \\
0 & \overline{A}_{11}^{s} & \overline{A}_{12}^{s} \\
\overline{A}_{21}^{m} & \overline{A}_{21}^{s} & \overline{A}_{22}^{o}
\end{bmatrix}
; \quad
\overline{x}^{ms} =
\begin{bmatrix}
\overline{x}^{m} \\
\overline{x}^{s} \\
\overline{x}^{o}
\end{bmatrix}
; \quad
\overline{b}^{ms} =
\begin{bmatrix}
\overline{b}^{m} \\
\overline{b}^{s} \\
\overline{b}^{o}
\end{bmatrix}
$$

and $\overline{x}^{m} = \left[i_{dc}^{m} \left(i_{1}^{m} \right)_{nh} \left(v_{s}^{m} \right)_{nh} \left(v_{o}^{m} \right)_{nh} \right]^{T};$ $\overline{x}^{s} = \left[i_{dc}^{s} \left(i_{1}^{s} \right)_{nh} \left(v_{s}^{s} \right)_{nh} \left(v_{o}^{s} \right)_{nh} \right]^{T};$ $\overline{x}^{o} = \left[\left(i_{o} \right)_{nh} \right]^{T};$
$\overline{A}_{11}^{m}, \overline{A}_{12}^{m}, \overline{A}_{21}^{m}, \overline{A}_{22}^{o}$ and $\overline{A}_{11}^{s}, \overline{A}_{12}^{s}, \overline{A}_{21}^{s}, \overline{A}_{22}^{o}$ are the partitioned submatrices of matrix equivalent to $\overline{A}$ of equation (xy), corresponding to the master and slave PAs, respectively; $\overline{b}^{m} = \overline{b}^{s} = \left[V_{dc} \left(0 \right)_{3nh} \right]^{T};$ and $\overline{b}^{o} = \left[\left(0 \right)_{nh} \right]^{T}.$ Note that elements of submatrices $\overline{A}_{11}^{m}, \overline{A}_{12}^{m}$ are constants only for fixed $\varepsilon_{i}^{m}, i = 0, 1, 2,$ and that of $\overline{A}_{11}^{s}, \overline{A}_{12}^{s}$ are constants for fixed δ^{s} and $\varepsilon_{i}^{s}, i = 0, 1, 2,$ and all other submatrices have fixed elements.

The flowchart of the iterative algorithm for the phase-controlled PA pair is given in Figure 7.2. The flowchart for the single-ended PA is an obvious subset of the flowchart and is not discussed here. Although the strategy for updating $\varepsilon_{i}^{mo}, \varepsilon_{i}^{so}, i = 0, 1, 2$ at each iteration can be chosen from known nonlinear solution methods (such as gradient descent techniques), a simple linear update law

$$
\begin{aligned}
\Delta\varepsilon_{i}^{mo} &= k_{\varepsilon}(\varepsilon_{i}^{m} - \varepsilon_{i}^{mo}) \\
\Delta\varepsilon_{i}^{so} &= k_{\varepsilon}(\varepsilon_{i}^{s} - \varepsilon_{i}^{so})
\end{aligned}
, \quad i = 0, 1, 2; \quad k_{\varepsilon} < 1 \tag{7.6}
$$

has been found adequate for the solution of most practical Class E–based amplifiers.

The higher the number of harmonics ($=n$) considered, the higher the accuracy of the solution; however, the number of resulting simultaneous equations and the variables to be solved increases with n, thereby increasing the computation complexity. For most practical PA designs, constraining n to 2 is a good tradeoff that facilitates high solution accuracy with low computation complexity.

The harmonic components of the circuit voltages and currents obtained at the end of the iteration process can be used to compute critical circuit variables such as the rms current and voltage stresses on the switching devices and other critical filter components. In addition, specific circuit waveforms can be reconstructed using the corresponding direct and quadrature axes components (which provide the amplitude and phase information), which may be useful in assessing other stress factors. For example, by reconstructing the switch voltage waveform, the presence of any hard switching of nonzero shunt capacitor voltages can be detected; moreover, the power losses due to such hard switching and the peak voltage impressed on the switching device can be computed.

7.2.1 Implementation of Harmonic Modeling

7.2.1.1 Single-Ended Class E Power Amplifier

The schematic of the example of the single-ended Class E PA, which is considered here to verify the performance of the harmonic model-based algorithm, is shown in

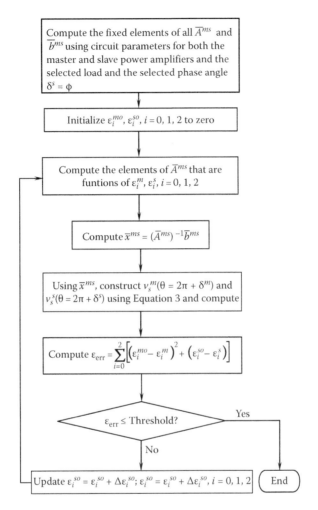

FIGURE 7.2   Flowchart of the harmonic modeling algorithm for phase-controlled Class E power.

Figure 7.2 [9]. The PA is rated for 500 W at 13.56 MHz and is optimized for a nominal (arbitrary) load of 145 Ω.

The harmonic model used in the algorithm uses only up to the second harmonic contents. The circuit parameters in Figure 7.3 are:

VDC1 = 150 or 175 V; L_{dc} = 3.9 mH; L_{i1} = 330 nH; L_{i2} = 390 nH; Lr = 330 nH; L_{o2} = 330 nH;

C_{i1} = 430 pF; C_{i2} = 82 pF; Cs = 328 pf; Cr = 1200 pF; Co1 = 100 pF; Co2 = 330 pF; Co3 = 330 pF;

Rci1 = 12 Ω and 0.15 mF; Variable Load

Figure 7.4 shows the comparison of the waveforms reconstructed from the results of the harmonic modeling of the example Class E PA with the corresponding

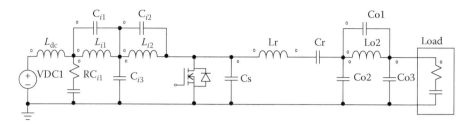

FIGURE 7.3 Schematic diagram of single-stage Class E power amplifier used for characterization examples.

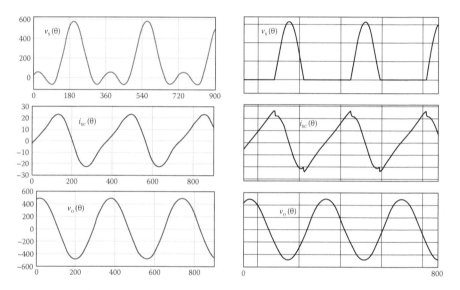

FIGURE 7.4 Comparison of power amplifier waveforms reconstructed from harmonic modeling with that from (a) harmonic model and (b) simulation.

waveforms obtained through PSIM simulation. The waveforms correspond to an RF control signal duty cycle of 0.5 and a resistive load with voltage standing wave ratio (VSWR) = 7.0. The close match of the two sets of waveforms and their amplitudes is obvious, with v_s and i_{sc} waveforms obtained from the algorithm clearly showing the characteristic second harmonic content.

A comparison of the results of the harmonic modeling with that of an experimental laboratory Class E PA is presented in Figure 7.5. The switch voltage waveforms are shown for multiple cases of rail voltage with the PA feeding a constant 50 Ω load and the RF control signal duty cycle held constant at 0.5.

Figure 7.5a shows the switch voltage waveforms of the experimental PA for six cases of rail voltage, and Figure 7.5b shows the corresponding waveforms obtained from harmonic modeling. The input and output power profiles obtained from the experimental setup and the modeling are compared in Figures 7.5c and d. The close

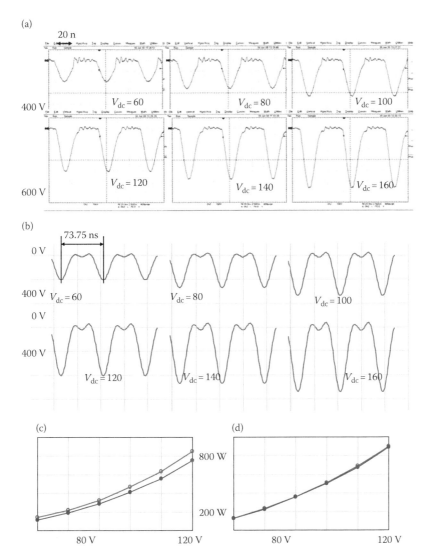

FIGURE 7.5 Comparison of switch voltage waveforms reconstructed from the harmonic modeling with experimental results, for various rail voltages with 50 Ω load. (a) Switch voltage waveforms of the experimental power amplifier, (b) Switch voltage waveforms reconstructed from harmonic modeling, (c) Input and output powers of the experimental power amplifier, and (d) Input and output powers of the power amplifier obtained through harmonic modeling.

qualitative match of the switch voltage waveforms between the harmonic model and the experimental amplifier is apparent from Figures 7.5a and b. The quantitative difference in the switch voltage values is obviously because of the fact that the power efficiency of the experimental amplifier is in the 78%–89% range for the rail voltage range considered, whereas the harmonic modeling does not account for conduction losses in the power switch and other components of the amplifier circuit.

Shown in Figure 7.6 is the comparison of the results of the harmonic-based algorithm with that of the simulation for the case of load variation through a constant VSWR circle in the Smith chart, with both the rail voltage V_{dc} and the duty cycle τ held constant at 150 V and 0.5, respectively, for all the runs. The close match of the performance of the harmonic model-based algorithm to that of the simulation is evident from all the variables depicted in the figure. The small errors seen in the rms values of the switch voltage and the nonconduction end angle $\varepsilon_0\,(=\varepsilon_1=\varepsilon_2$, for this case) can be attributed to neglecting the higher-order harmonic components above the second in the harmonic model-based algorithm. Also evident from Figure 7.5 is the classic characteristic of Class E (or any switch-mode) PA reported in the literature (e.g., [10]), where the output power steeply rises in the

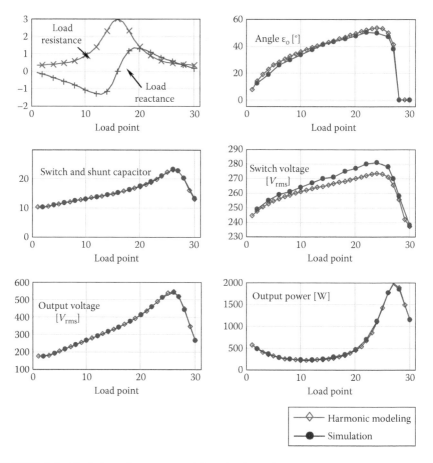

FIGURE 7.6 Variation of power amplifier circuit variables with load (VSWR = 3.0, V_{dc} = 150 V, τ = 0.5).

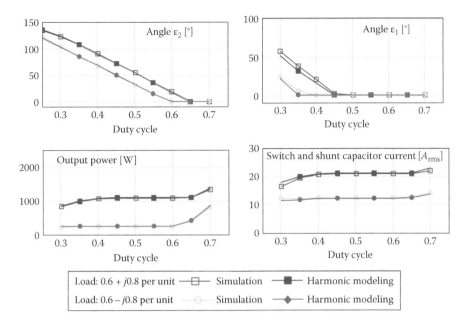

FIGURE 7.7 Variation of power amplifier circuit variables with duty cycle of the radio frequency gate signal ($V_{dc} = 150\,\text{V}$).

neighborhood of the low-resistance load point at which the load reactive component is in near resonance with the equivalent output reactance of the PA. It is well known that this characteristic warrants the need for additional protection schemes such as inductive clamping [11].

Figure 7.7 shows the comparison between the harmonic model-based algorithm and the circuit simulation in terms of PA circuit variables as functions of the duty cycle τ, for two cases of load points, a resistor–capacitor and a resistor–inductor. The results match up closely for all variables including the angles ε_1 and ε_2. The errors noted in the case of ε_1 are obviously due to the difficulty in accurately modeling the discontinuities of the switch voltage waveform using only limited harmonic order components ($n = 2$).

Figure 7.8 shows the characteristics of the Class E PA with the adjustment of the DC rail voltage, V_{dc}, derived using the harmonic modeling. Results are shown in Smith charts for the load variation through a constant VSWR circle, for three constant VSWR cases—1.5, 3.0, and 5.0. From Figure 7.6b, the basic nonlinearity of the required rail voltage profile to maintain constant output power with reference to the load point within a constant VSWR circle is apparent from the offset of the three loop curves from the center. More importantly, the voltage rail curves for the higher VSWR loads show that when the load has low resistance with the reactive component in close resonance with the output impedance of the PA, the voltage rail should be substantially lowered to regulate the output power at the desired value. Figure 7.8 also indicates that the rms voltage across the switch is in approximate proportion with the rail voltage for the various load conditions (as

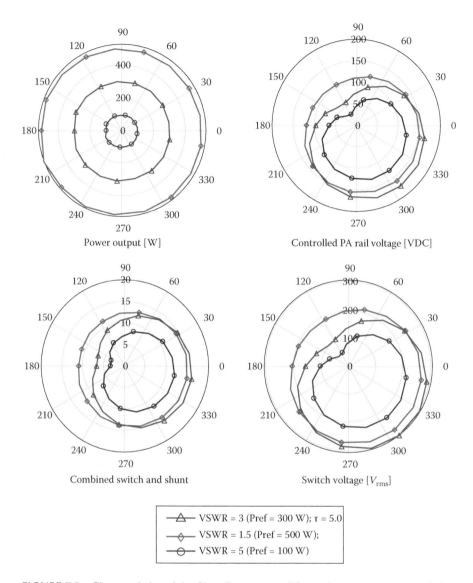

FIGURE 7.8 Characteristics of the Class E power amplifier under output power regulation through rail voltage control, for constant VSWR loads, $\tau = 0.5$.

seen from the similar shape and angular orientation of the corresponding curves in the Smith chart). However, the rms value of the combined current through the switch and the capacitor has a proportional, but phase-shifted, relationship with the rail voltage (although the voltage and current curves have similar shape, the current curves have a leading angular orientation with reference to the voltage curves in the Smith chart). The noted phase shift between the voltage and current profiles is due to the impact of the filtering circuits of the Class E amplifier and their reactive values relative to the load.

7.2.1.2   Phase-Controlled Class E Power Amplifier

In this section, we consider a phase-controlled PA pair made up of two identical stages of the single-ended Class E PA shown in Figure 7.3, connected through an ideal 1:1 combiner-transformer. For the purpose of normalization, and to allow comparative analysis with the single-ended amplifier, the load values considered for the PA pair are selected twice that of the corresponding load values of the single-ended amplifier.

A comparison of the results of the harmonic modeling of the phase-controlled PA pair with that of the corresponding experimental implementation is presented in Figure 7.9. The switch voltage waveforms of the master and the slave PAs of the experimental PA pair, and the corresponding waveforms reconstructed from the harmonic model shown for a sample case of load, phase control angle, and rail voltage, indicate that the results of the harmonic modeling closely match that of the experimental implementation.

Figure 7.10 depicts the output power profile (z-axis) as a function of the phase control angle (x-axis) with reference to variation in the load impedance through a constant VSWR circle (y-axis). It indicates that the output power is controllable in a near-linear fashion through phase control, for a given load condition. Figure 7.11 shows the DC input power (z-axis) drawn by each PA as a function of the phase control angle (x-axis) with reference to variation in the load impedance through a constant VSWR circle (y-axis). It indicates that the DC input power imbalance between the master and slave PAs is minimum at the extremes of the phase angle range (i.e., near 180° and 0°), and in some range, part of the input power drawn by the master PA is returned to the DC source through the slave PA, as seen from the negative values of the DC input current of the slave amplifier.

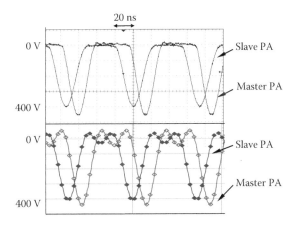

FIGURE 7.9   Comparison of switch voltage waveforms of the phase-controlled Class E power amplifier pair reconstructed from harmonic modeling with the experimental results – Load: $40 - j30\ \Omega$; $V_{dc} = 140\,V$; Output power = 350 W.

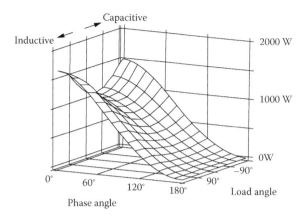

FIGURE 7.10 Output power (z-axis) vs phase angle (y-axis) for load points around VSWR=1.25 circle (x-axis).

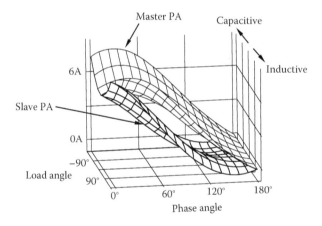

FIGURE 7.11 DC input current (z-axis) of the master and slave PAs vs. phase angle (y-axis) for load points around VSWR=1.25 circle (x-axis).

Figure 7.12 shows the characteristics of the Class E PA pair with output power regulated through the adjustment of the phase angle of the RF signal of the slave PA, derived using the harmonic modeling, with the results shown in Smith charts for the load variation through three constant VSWR circles. The following are apparent from the figure: The combined rms current through the switch and the shunt capacitor is higher in the master PA than in the slave amplifier for all loading conditions, whereas the rms voltage across the switch in both the master and the slave amplifiers is about the same. Also from Figures 7.8 and 7.12, when compared with the single-ended PA where the power regulation is achieved through rail voltage control, the combined switch-capacitor current and the switch voltage are significantly higher for the phase-controlled amplifier pair, for the same per-amplifier output power. This indicates that the rail voltage control as well as the phase control may have significant

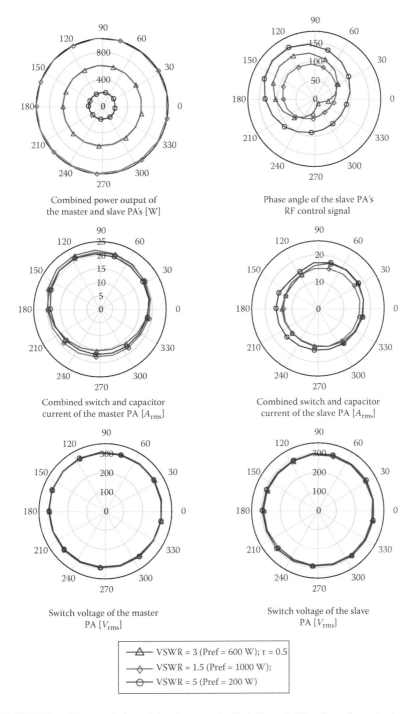

Combined power output of
the master and slave PA's [W]

Phase angle of the slave PA's
RF control signal

Combined switch and capacitor
current of the master PA [A_{rms}]

Combined switch and capacitor
current of the slave PA [A_{rms}]

Switch voltage of the master
PA [V_{rms}]

Switch voltage of the slave
PA [V_{rms}]

VSWR = 3 (Pref = 600 W); τ = 0.5
VSWR = 1.5 (Pref = 1000 W);
VSWR = 5 (Pref = 200 W)

FIGURE 7.12 Characteristics of the phase-controlled Class E PA pair under output power regulation, for constant VSWR loads, τ = 0.5.

roles to play in practical PA control designs: For example, where high-speed output transient response is important, a fast-control loop can be designed with the phase angle as the primary control mechanism; on the other hand, high-speed response can be traded off for high efficiency and/or low voltage and current stresses using the rail voltage control as the main control method.

Overall, the systematic development of a harmonic model-based analysis tool facilitates the characterization of Class E–based amplifiers for arbitrary loading conditions. The analysis exploits the fact that the nonlinearity of the basic Class E amplifier is confined to the single-ended switching action, and that the resulting voltage and current signals have dominant harmonics of specific known order (e.g., $n = 2$). The use of both the direct and the quadrature components of the fundamental and harmonics up to a selected order allows the voltage and current equations of the PA to be established in a quasi-linear form, which are then solved using conventional iterative solution methods. The analysis algorithm is extended to characterize the phase-controlled Class E PA pair, using direct augmentation of circuit equations. Comparison with simulation and representative experimental results shows that the characterization through harmonic modeling has very good and acceptable accuracy levels. Higher accuracy levels can be achieved by including higher-order harmonic contents in the modeling, however, at the cost of increased computation complexity. Characterization through harmonic modeling is highly useful in the circuit design process of Class E-based PAs, for example, in rapidly assessing the voltage and current stresses on switching devices and critical filtering components, and for the comparative impact analysis of competing control methodologies, all without the need for extensive circuit simulation tools.

7.3 PULSING OF SWITCH-MODE AMPLIFIERS

The RF envelope pulsing technique using phasing with harmonic modeling to pulse switch-mode amplifiers have been introduced by this author and his colleague in Reference [12] and implemented for Class E amplifiers. In this section, pulsing of Class E amplifiers will be discussed, and formulation, simulation, and measurement results will be illustrated.

7.3.1 PULSING OF CLASS E AMPLIFIERS

The harmonic modeling of phase-controlled Class E amplifier is detailed in References [9,12] and can be illustrated in Figure 7.13. The phase-controlled amplifier shown in Figure 7.13 consists of two single-ended amplifiers called Master PA and Slave PA. Ideal switches and components are used to demonstrate the operational principle of the amplifier and derive the equations. In Figure 7.13, the average of the current through the transistor Q_1 can be written as

$$i_{so} = \frac{1}{2\pi} \int_{\delta}^{\psi+\delta} i_{sc}(\theta) \, d\theta \tag{7.7}$$

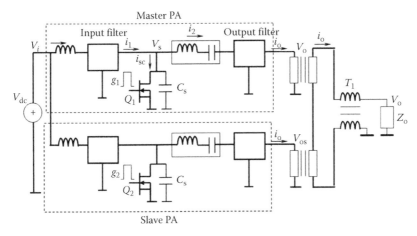

FIGURE 7.13   Illustration of phase-controlled Class E amplifiers.

Q_1, which is also considered to be an ideal switch, is turned on at an angle δ, and it conducts between the angles δ and $\psi + \delta$. i_{sc} is the steady-state current and defined by

$$i_{sc}(\theta) = i_{dc} + \sum_{h=1}^{n} [(i_{1dh} - i_{2dh}) \sin (h\theta) + (i_{1qh} - i_{2qh}) \cos (h\theta)] \qquad (7.8)$$

Current in Equation 7.8, i_{sc}, is the current on the master side of the Class E amplifier as shown in Figure 7.13 and has fundamental and harmonic components. i_{dh} and i_{qh} are the direct and quadrature harmonic components of the current, respectively, and h is used to define the number of the harmonics. The gate drive signals, g_1 and g_2, on master and slave of the amplifier, are basically phased to produce the pulsed output.

If the master and slave PAs are phased by 180°, the ON/OFF pulsing is obtained, and if they are phased with any other angle, the two-level pulsing is obtained. When the phase difference is 180°, the maximum pulsed output power is delivered. The gate drive signals, g_1 and g_2, which are used to excite gates of the transistors of master and slave PAs are shown in Figure 7.14. The theoretical profiles for power, current, and voltage for the phase-controlled Class E amplifiers are obtained using Mathcad with the formulation presented in References [9,12]. The power profile of the phase-controlled Class E amplifier for inductive and capacitive loads versus phase angle is shown in Figure 7.15. The power profile shows that there is an imbalance in DC input power between the master and slave PAs, and this imbalance is minimum 180° and 0°.

The rms voltage and current profile of the transistors for phase-controlled Class E amplifiers are illustrated in Figure 7.16a and b, respectively. Based on the analytical results illustrated in Figure 7.16a, the voltage across the transistors significantly increases when the phase approaches 0°. It is concluded from Figure 7.16b that the current imbalance between the master and the slave is worse when the phase difference is between 0° and 180°.

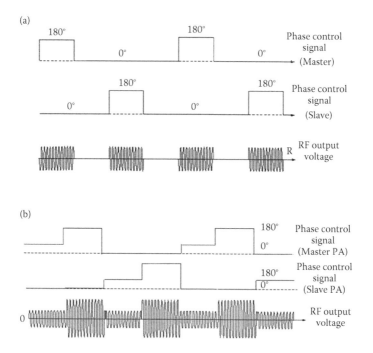

FIGURE 7.14   Illustration of ON/OFF and two-level pulsing of Class E amplifiers (a) ON/OF pulsing of Class E amplifiers and (b) two-level pulsing of Class E amplifiers.

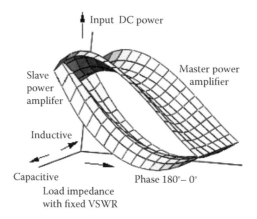

FIGURE 7.15   Analytical power profile of phase-controlled Class E amplifiers.

7.3.1.1   Simulation Results

The phase-controlled Class E amplifier for pulsing applications is simulated, and the amplifier is tested for several load conditions during pulsing. The design specifications for the phase-controlled Class E amplifiers are given in Table 7.1.

Phase-controlled amplifier consists of two single-ended Class E amplifiers that are connected with output balun. Each of the single-ended amplifier is fed via RF

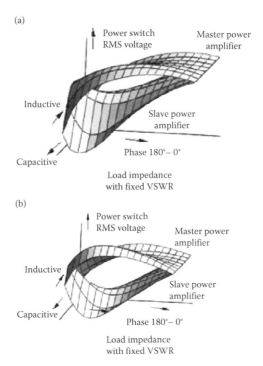

FIGURE 7.16   Analytical (a) voltage and (b) current profile of the transistors for phase-controlled Class E amplifiers.

TABLE 7.1

Specifications for Phase-Controlled Class E Amplifier

CW RF Frequency (MHz)	Pulse Frequency Range (kHz)	Rated CW Power (W)	Agile V_{dc} (V)	Rated Pulsed Power (W)
13.56	0–100	1800	0–180	500

CW, continuous wave; RF, radio frequency.

pulse signal that is phased based on the desired power level as shown in Figure 7.14. The schematic of the simulated and experimented phase-controlled Class E amplifier for pulsing is shown in Figure 7.17. Nonlinear time domain circuit simulator, Pspice, is used to characterize the pulsing amplifier.

In the phase-controlled pulsing amplifier circuit, IXFH12N100 Metal Oxide Field Effect Transistors (MOSFETs) are connected in parallel for the master and slave sides. Each side of the amplifier provides approximately 900 W RF power to a matched load when rail voltage is set to 180 V in continuous wave (CW) mode. Power is doubled when they are combined via output balun as shown in Figure 7.17. MOSFETs are rated for 1000 V and 12 A, and as a result, it is important not to exceed

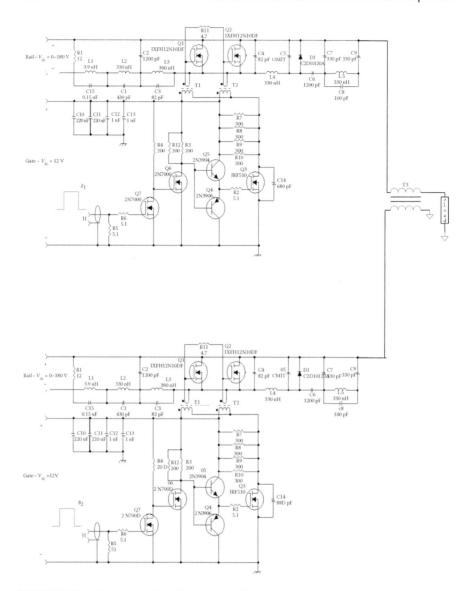

FIGURE 7.17   Phase-controlled Class E amplifier for pulsing applications.

these rated values during pulsing. The simulation results when pulsing to a matched load with 50% duty cycle, $D = 0.5$, and 67.8 kHz pulse period, PP = 67.8 kHz at $V_{dc} = 180$ V are illustrated in Figure 7.18. As seen, the ideal pulse is used to drive the MOSFETs, and the maximum output is obtained when master and slave PAs are out of phase by 180°, and minimum power is obtained when they are in phase.

The average power during pulsing ranges from 892 to 1033 W as shown in Figure 7.18. The power and drain-to-source impedance distribution versus phase difference between the master and slave PAs are also simulated and plotted on the Smith

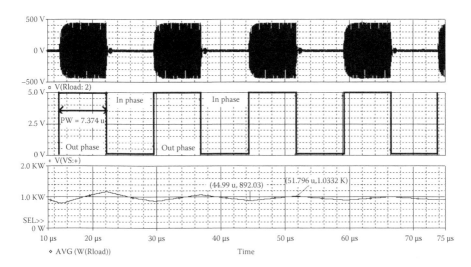

FIGURE 7.18 Simulated pulse waveforms for a matched load with $D = .0.5$ and PP = 67.8 kHz at $V_{dc} = 180$ V.

chart (Figure 7.19). Based on the simulation results, power distribution is maximized when the phase difference between the master and slave PAs is 180°. The drain-to-source impedance distribution for the master and slave PAs versus phase is different as shown in Figure 7.19a and b, respectively. Hence, this results in a variation of the power distribution provided by master and slave PAs. The results confirm the ones obtained in Figure 7.15. In addition, the drain-to-source voltage for master and slave PAs are different due to the impedance variation.

This is illustrated in Figure 7.20a and b, respectively. Master and slave PAs experience higher voltage levels as the phase approaches zero, as confirmed with the analytical results in Figure 7.16a. The simulated voltage values are below the breakdown voltages of the transistors. Each transistor's (die) dissipation for proper operation is defined by the MOSFET manufacturer to be 300 W. This determines the thermal profile of the transistors used in the amplifier. The die dissipation or thermal profile of the phase-controlled Class E amplifier versus phase is simulated and shown in Figure 7.21. The dissipation simulated was always less than 110 W. This value is much lower than the maximum dissipation level. It is important to note also that the die dissipation was worse for the transistors of the slave PA in comparison to the die dissipation of the transistors for master PA. Class E phase-controlled RF amplifier is also simulated and pulsed for high VSWR loads. The simulated impedance points when load has VSWR of 2 is shown in Figure 7.22. The pulse waveforms when amplifier is pulsed with the same condition described in Figure 7.18 are illustrated in Figure 7.23. The average power during pulsing for the impedance point 2 in Figure 7.21 ranges from 335 to 418 W. RF pulsed waveforms are also observed to sustain their characteristics during pulsing.

After simulating and pulsing the phase-controlled Class E amplifier for several cases, it has been confirmed that the simulated values were always within the safe

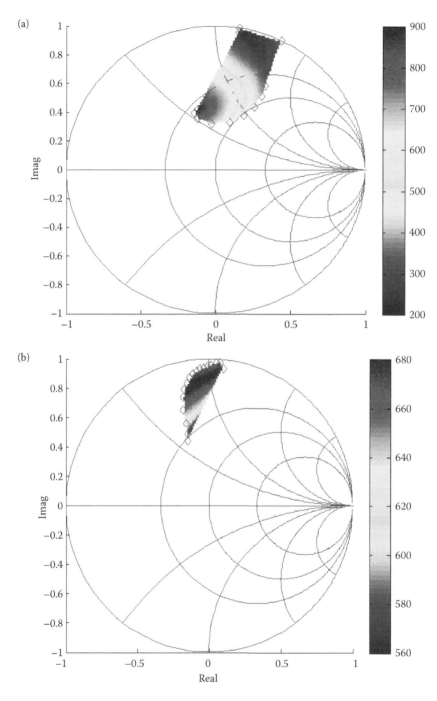

FIGURE 7.19 Simulated power distribution and drain-to-source impedance vs distribution and driving gate signals phase difference. (a) Master PA and (b) Slave PA.

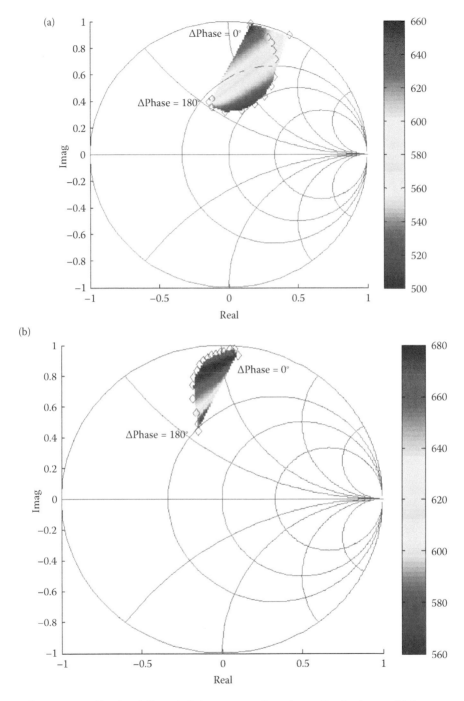

FIGURE 7.20 Simulated V_{ds} and drain-to-source impedance distribution vs driving gate signals phase difference. (a) Master PA and (b) Slave PA.

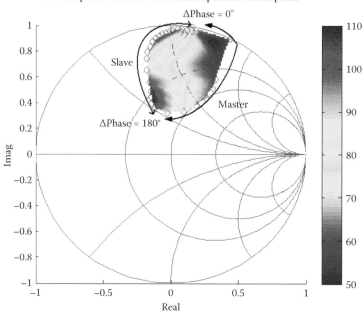

FIGURE 7.21 Simulated die dissipation for phase-controlled pulsing PA.

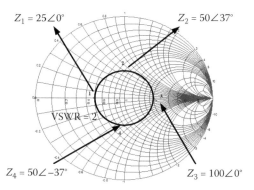

FIGURE 7.22 The impedance points that amplifier is pulsed when VSWR = 2.

operating area (SOA) of the components, and the amplifier was able to produce the required RF pulse waveforms. Hence, the phase-controlled PA is built and tested. The measurement results are detailed in Section 7.3.1.2.

7.3.1.2 Experimental Results

The phase-controlled pulsing Class E amplifier in Figure 7.13 is built and tested. The constructed and tested amplifier is shown in Figure 7.24. The amplifier is tested for various pulsing frequencies for matched and mismatched loads and characterized.

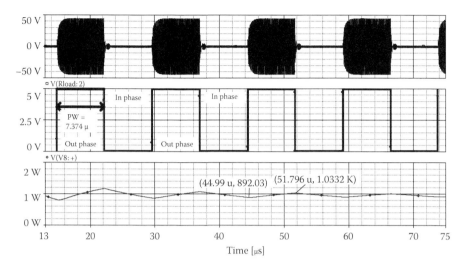

FIGURE 7.23 Simulated pulse waveforms $Z = 50\angle37°$ with VSWR = 2 and $D = 0.5$ and PP = 67.8 kHz at $V_{dc} = 140$ V.

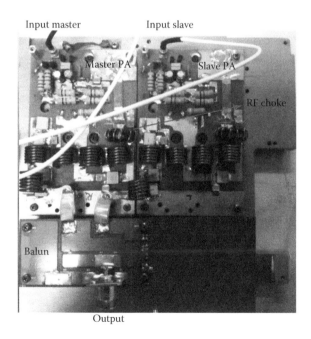

FIGURE 7.24 Constructed and tested phase-controlled pulsing Class E amplifier.

7.3.1.2.1 Matched Load Characteristics of Pulsing Class E Amplifier

The drain efficiency for a matched load when pulsing frequencies range from 20 to 100 kHz is given in Figure 7.25. The duty cycle is set to be $D = 0.5$, and rail voltage is adjusted to 140 V_{dc} during the testing.

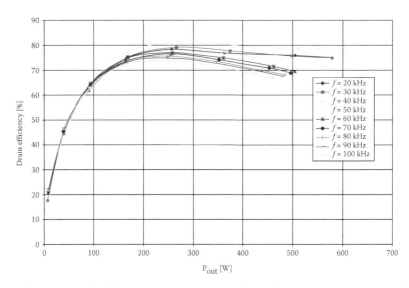

FIGURE 7.25 Drain efficiency vs output power for various pulsing frequencies for a matched load.

The drain efficiency is lowest at the highest pulsed frequency. Die dissipation for pulsing frequencies from 20 to 100 kHz are also measured and illustrated in Figure 7.26. The die dissipation is worst when the amplifier is pulsed with 100 kHz pulsed frequency. However, the die dissipation is well within the SOA for all cases when the amplifier is terminated to a matched load. The measured rise time, fall

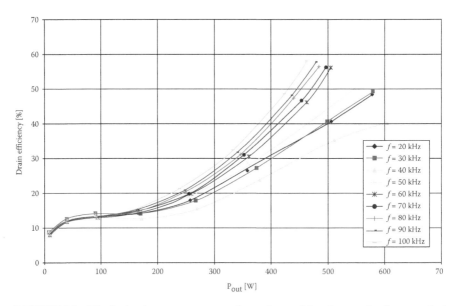

FIGURE 7.26 Die dissipation vs output power for various pulsing frequencies for a matched load.

time, settling time, and delay time for phase-controlled pulsing Class E amplifier versus various pulse frequencies when amplifier was pulsing to a matched load are illustrated in Figures 7.27 through 7.30, respectively.

The rise time varies from 1.47 to 0.93 μs, whereas fall time changes from 0.36 to 0.43 μs as pulse frequency is varied from 20 to 100 kHz. Settling time stays between 2.2 and 2.6 μs, and delay time varies from 1.6 to 2.1 μs as pulse frequency changes. The variation in the drain-to-source voltage of the transistors during pulsing is also measured and shown in Figure 7.31. It is seen that the drain-to-source voltage is

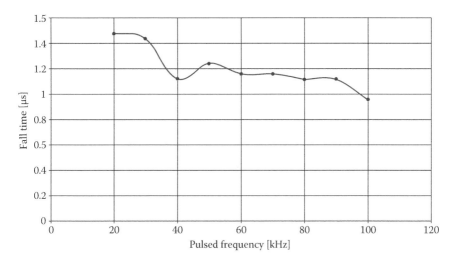

FIGURE 7.27 Rise time vs pulsing frequency phase-controlled Class E pulsing amplifier.

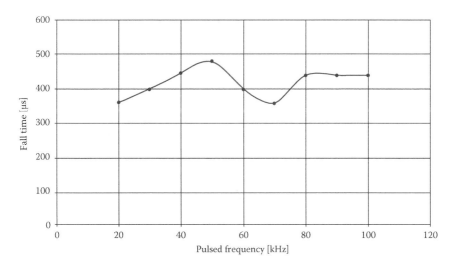

FIGURE 7.28 Fall time vs various pulse frequencies for phase-controlled Class E pulsing amplifier.

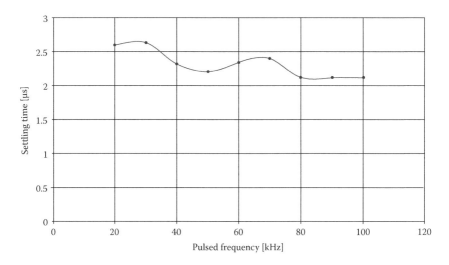

FIGURE 7.29   Settling time vs various pulse frequencies for phase-controlled Class E pulsing amplifier.

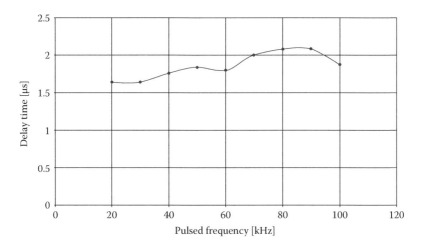

FIGURE 7.30   Delay time vs various pulse frequencies for phase-controlled Class E pulsing amplifier.

between 887 and 914 V when the amplifier is pulsed at the preset conditions for the pulse frequencies from 20 to 100 kHz.

Class E phase-controlled amplifier pulse waveforms are obtained when amplifier is pulsed from 20 kHz to 100 kHz for ON/OFF pulsing, where master and slave PAs are out-phased by 180° (Figure 7.32). It is observed that the amplifier was able to maintain its stability during pulsing even for the high-pulse frequencies. In Figures 7.32 through 7.34, the zoomed-in RF pulse envelope waveforms are given for reference for each pulse frequency.

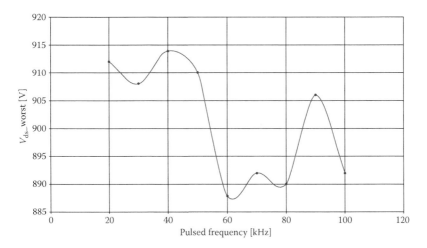

FIGURE 7.31 Transistor drain-to-source voltage, V_{ds}, vs pulse frequencies from 20 to 100 kHz when phase-controlled Class E amplifier is terminated with a matched load.

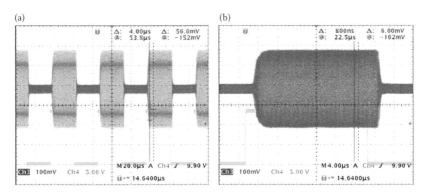

FIGURE 7.32 Phase-controlled Class E amplifier pulse waveforms for ON/OFF pulsing when (a) pulse frequency (PF) = 20 kHz and (b) Zoomed in.

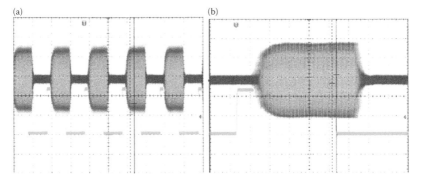

FIGURE 7.33 Phase-controlled Class E amplifier pulse waveforms for ON/OFF pulsing when (a) pulse frequency (PF) = 50 kHz and (b) Zoomed in.

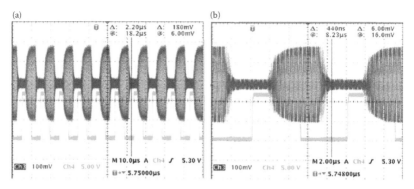

FIGURE 7.34 Phase-controlled Class E amplifier pulse waveforms for ON/OFF pulsing when (a) pulse frequency (PF) =100 kHz and (b) Zoomed in.

In practice, plasma chambers might be required to be energized with different pulse frequencies and power levels for some processes. This requires amplifiers to have the capacity of two-level pulsing without any stability problems. Hence, the phase-controlled Class E RF amplifier is also tested for two-level pulsing, when it is terminated with matched load for different pulse frequencies. The same operating conditions have been used by setting duty cycle to $D=0.5$ and rail voltage to 140 V_{dc}, and out phasing of master and slave PAs. The pulse waveforms when the amplifier is pulsed with pulse frequencies between 20 and 100 kHz are illustrated in Figures 7.35 through 7.37. It is important to note that Class E amplifier that is designed based on harmonic modeling was able to provide the required two-level pulsing power levels with no oscillation as desired.

7.3.1.2.2 Mismatched Load Characteristics of Pulsing Class E Amplifier

Class E phase-controlled amplifier shown in Figure 7.24 is terminated with various high VSWR loads to have complete characterization. RF amplifier was terminated with the load impedance points shown in Figure 7.22 and pulsed at different frequencies. The response of the amplifier for drain efficiency and die dissipation versus output power are given in Figures 7.38 and 7.39, respectively. As seen from the figures,

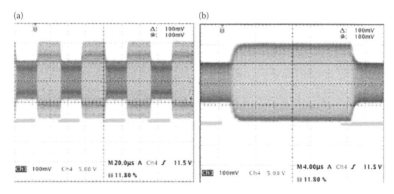

FIGURE 7.35 Phase-controlled Class E amplifier pulse waveforms for two-level pulsing when (a) pulse frequency (PF)=20 kHz and (b) Zoomed in.

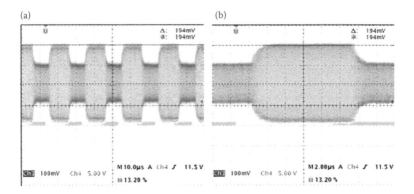

FIGURE 7.36 Phase-controlled Class E amplifier pulse waveforms for two-level pulsing when (a) pulse frequency (PF)=50 kHz and (b) Zoomed in.

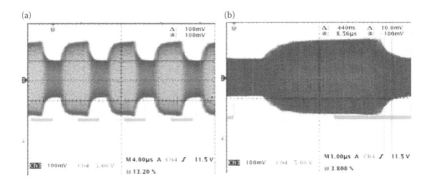

FIGURE 7.37 Phase-controlled Class E amplifier pulse waveforms for two-level pulsing when (a) pulse frequency (PF)=100 kHz and (b) Zoomed in.

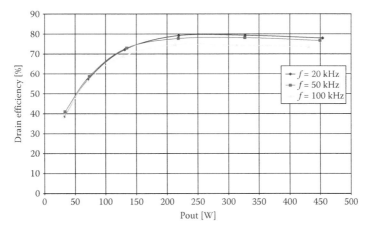

FIGURE 7.38 Drain efficiency vs output power for various pulsing frequencies for $Z=50\angle 37°$ with VSWR=2 and $D=0.5$ and PP=67.8 kHz at $V_{dc}=140$ V.

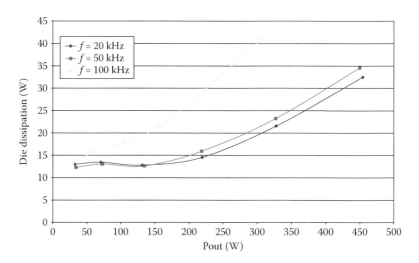

FIGURE 7.39 Die dissipation vs output power for various pulsing frequencies for $Z=50\angle37°$ with VSWR$=2$ and $D=0.5$ and PP$=67.8$ kHz at $V_{dc}=140$ V.

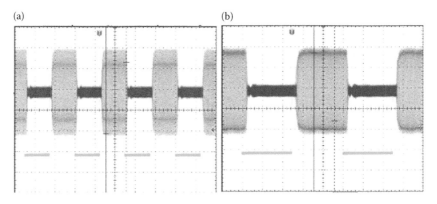

FIGURE 7.40 Phase-controlled Class E amplifier pulse waveforms for ON/OFF pulsing for a mismatched load with PF$=20$ kHz and $D=0.5$ when (a) $Z=25\angle0°$ and (b) $Z=25\angle0°$ zoomed in.

the change in drain efficiency is minimal, and die dissipation remains to be less than 45 W for pulse frequencies from 20 to 100 kHz.

Class E phase-controlled amplifier is also pulsed at four different impedance points with VSWR of 2 as shown in Figure 7.22 when PF$=20$ kHz, $D=0.5$, and rail voltage$=140$ V_{dc}. The pulse waveforms are illustrated in Figures 7.40 through 7.43.

As shown, RF amplifier designed and pulsed using phasing based on harmonic modeling was always able to keep its stability. The overshoot is also measured to be less than 5%. The stability of the amplifier is measured with spectrum analyzer for the mismatched loads (Figure 7.44). The spurious and harmonic levels are less than −64 dB as shown in the figure.

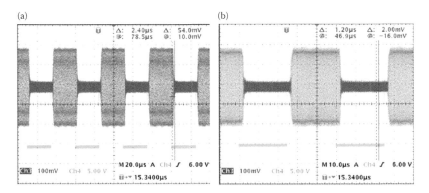

FIGURE 7.41 Phase-controlled Class E amplifier pulse waveforms for ON/OFF pulsing for a mismatched load with PF=20kHz and $D=0.5$ when (a) $Z=50\angle37°$ and (b) $Z=50\angle37°$ zoomed in.

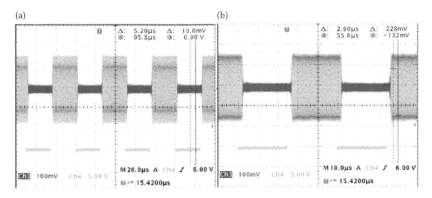

FIGURE 7.42 Phase-controlled Class E amplifier pulse waveforms for ON/OFF pulsing for a mismatched load with PF=20kHz and $D=0.5$ when (a) $Z=100\angle0°$ and (b) $Z=100\angle0°$ zoomed in.

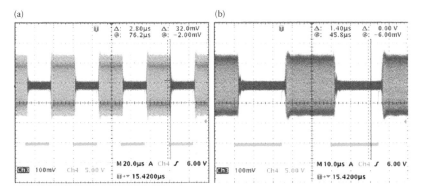

FIGURE 7.43 Phase-controlled Class E amplifier pulse waveforms for ON/OFF pulsing for a mismatched load with PF=20kHz and $D=0.5$ when (a) $Z=50\angle-37°$ and (b) $Z=50\angle-37°$ zoomed in.

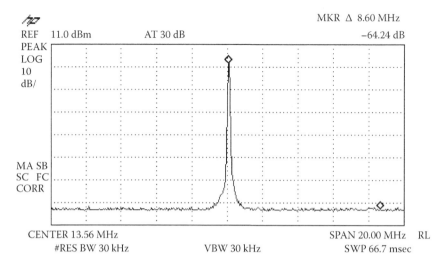

FIGURE 7.44 Spectral content of phase-controlled Class E amplifier when pulsed with a mismatched load with VSWR = 2.

REFERENCES

1. C.Q. Hu, X.Z. Zhang, S.P. Huang, Class-E combined converter by phase shift control, *1989 IEEE Power Electronics Specialists Conference*, pp. 229–243, 1989.
2. K. Shinoda, T. Suetsugu, M. Matsuo, S. Mori, Analysis of phase-controlled resonant DC-AC inverters with Class-E amplifier and frequency multipliers, *IEEE Transactions on Industrial Electronics*, Vol. 45, No. 3, pp. 412–420, 1998.
3. I. Boonyaroonate, S. Mori, Class E phase-controlled inverter with single RF choke, *20th International Telecommunications Energy Conference (INTELEC)*, San Francisco, pp. 549–553, October 1998.
4. K. Kawamoto, H. Sekiya, K. Koizumi, I. Sasase, Design of a generalized phase-controlled Class-E inverter, *The 2001 IEEE International Symposium on Circuits and Systems*, Sydney, Australia, pp. 393–396, May 2001.
5. H. Sekiya, J. Lu, T. Yahagi, Phase control for resonant DC-DC converter with Class-DE inverter and Class-E rectifier, *IEEE Transactions on Circuits and Systems, Fundamental Theory and Applications*, Vol. 53, No. 2, pp. 254–263, 2006.
6. K. Tom, M. Faulkner, T. Lejon, Performance analysis of pulse width modulated RF Class-E amplifier, *63rd IEEE Vehicular Technology Conference*, Melbourne, Australia, pp. 1807–1811, May 2006.
7. K. Tom, V. Bassoo, M. Faulkner, T. Lejon, Load-pull analysis of outphasing Class-E power amplifier, *2nd International Conference on Ultra Wideband Communication*, Sydney, Australia, pp. 52–56, August 2007.
8. H. Sekiya, I. Sasase, S. Mori, Computation of design values of Class E amplifiers without using waveform equations, *IEEE Transactions on Circuits and Systems I: Fundamental Theory and Applications*, Vol. 49, No. 1, pp. 966–978, 2002.
9. S. Sivakumar, A. Eroglu, Analysis of Class-E based RF power amplifiers using harmonic modeling, *IEEE Transactions on Circuits and Systems I: Regular Papers*, Vol. 57, No. 1, pp. 299–311, 2010.

10. R. Frey, 500 W, Class-E 27.12 MHz amplifier using a single plastic MOSFET, *1999 IEEE MTT-S International Microwave Symposium Digest*, Vol. 1, pp. 359–362, Anaheim, CA, June 1999.
11. D. Lincoln, P. Bennett, Class E amplifier with inductive clamp, U.S. Patent # 7180758, February 20, 2007.
12. S. Sivakumar, A. Eroglu, Radio frequency (RF) envelope pulsing using phase switching of switch-mode power amplifiers, US Patent: US 7872523 B2, 2011.

8 Distortion and Modulation Effects in RF Power Amplifiers

8.1 INTRODUCTION

In transceiver systems, misalignments in the quadrature modulator feeding a nonlinear high-power amplifier can produce distortion products that lead to spectrum leakage into adjacent channels. In this chapter, elimination of these distortion products will be discussed and detailed.

Quadrature modulation is a convenient and flexible way to produce nearly any type of waveform and has been in use for decades. As such, quadrature modulators are ubiquitous in communication systems as shown in Figure 8.1, and the performance of this device is important to the overall system [1]. Ideal quadrature modulators generate single continuous wave (CW) tone that can be used in radio frequency (RF) systems. The ideal system assumes no gain or phase changes in either the phase (I) or quadrature (Q) paths as shown in Figure 8.1. However, with real components, there will be losses, phase changes, and DC offsets that need to be taken into account to have a balanced system. IQ imbalance reduces the dynamic range and as a result degrades the performance of the system [2].

The effect of IQ imbalance on RF systems has been investigated by several researchers [3–8]. The linear imperfection feature effects of IQ gain and phase imbalance on image rejection ratio and error vector magnitude have been presented in References [3–4].

The effect of IQ imbalances on orthogonal frequency division multiplexing (OFDM) has been studied in References [5–7]. The effect of I/Q imbalance in the front-end section of low intermediate frequency receiver has been investigated in Reference. [8]. Several methods to improve IQ imbalance have been presented and discussed in the literature [9–16]. In References [9–13], lengthy and time-consuming compensation algorithms for OFDM receivers have been proposed to balance IQ paths. Adaptive equalization techniques using known preambles to balance IQ modulators are given in References [14–15].

The *K*-matrix modeling of IQ imbalance is presented in Reference. [16]. This requires extraction of *K*-matrix parameters using large signal network analyzer measurements [17,18]. There are several other similar methods presented to improve I/Q imbalance in the literature [19–28]. As a result, the proposed methods to improve IQ imbalance in the literature are most of the time impractical, invasive, require calibration of the system before its operation, costly, and rely on several measurements that are used for precalibration and stored in the memory.

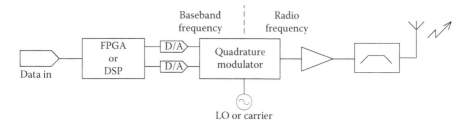

FIGURE 8.1   Typical transmitter system.

In this chapter, a noninvasive IQ balancing system based on the image band rejection method is introduced to eliminate the distortion effects seen in RF amplifiers due to imperfections in IQ modulators when they are used as an RF input source [19]. The balancing system is furthermore simulated with nonlinear circuit simulator Advance Design System (ADS) for verification. The implementation of the proposed IQ balancing system is detailed step by step and shown in ADS. The analytical model is compared with the simulation results. It has been shown that drastic improvement has been obtained in the performance of the quadrature modulator when the method proposed is implemented. The method introduced in this chapter is noninvasive, which requires mixing RF output of the quadrature modulator with a signal at the image frequency that is phase-locked with the test signal. In the system discussed, unlike all other existing methods, only two sets of measurements during the operation are needed to obtain the correction factors. The correction factors are then used for self-calibration and optimization of image band signal, which makes the system noninvasive. A practical and simple IQ balancing method for nonideal quadrature modulators such as the one introduced here is applicable in RF systems where cost, time, and accuracy are important factors.

8.2   IDEAL QUADRATURE MODULATOR

Figure 8.2 shows the block diagram of an ideal quadrature modulator. In practice, the I and Q signals could be generated digitally from a Field-Programmable Gate Array (FPGA) device and Digital-to-Analog Converter (DAC) combination. Low-pass filters would also be necessary to eliminate the aliasing effect of the digitally generated I and Q signals. The local oscillator (LO) could be created from a phase-locked loop (PLL) or some other tunable CW source. The signal at the output of the mixer on I and Q paths can be expressed as

$$I_{up} = \frac{1}{2}\Big[\cos\big(\omega_{BB}t + \omega_{LO}t\big) + \cos\big(\omega_{BB}t - \omega_{LO}t\big)\Big] \tag{8.1}$$

$$Q_{up} = \pm\frac{1}{2}\Big[\cos\big(\omega_{BB}t - \omega_{LO}t\big) - \cos\big(\omega_{BB}t + \omega_{LO}t\big)\Big] \tag{8.2}$$

The output signal from the modulator is then obtained from

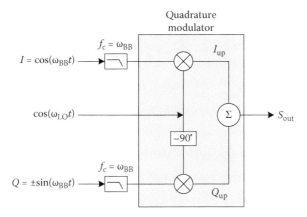

FIGURE 8.2   Ideal quadrature modulator.

$$I_{up} + Q_{up} = \frac{1}{2}\left[\cos\left(\omega_{BB}t + \omega_{LO}t\right) + \cos\left(\omega_{BB}t - \omega_{LO}t\right)\right]$$

$$\pm \frac{1}{2}\left[\cos\left(\omega_{BB}t - \omega_{LO}t\right) - \cos\left(\omega_{BB}t + \omega_{LO}t\right)\right] \qquad (8.3)$$

From Equation 8.3, it is seen that by changing the sign of the Q input signal, either the upper sideband $(\omega_{BB}t + \omega_{LO}t)$ or the lower sideband $(\omega_{BB}t + \omega_{LO}t)$ can be cancelled. This can be illustrated in Figure 8.3. For instance, when $Q = \sin\left(\omega_{BB}t\right)$ as shown in Figure 8.3a, then we have perfect cancellation, or quadrature nulling of the upper sideband, and as a result, we obtain the ideal modulator output signal as

$$S_{out} = \cos\left(\omega_{BB}t - \omega_{LO}t\right) \qquad (8.4)$$

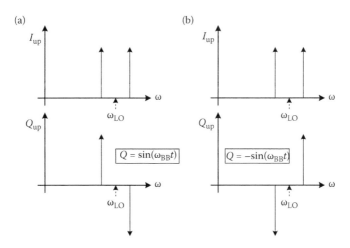

FIGURE 8.3   Ideal quadrature modulator summation when (a) $Q = \sin(\omega_{BB}t)$ (b) $Q = -\sin(\omega_{BB}t)$.

In physically realizable systems, the antialiasing filters do not necessarily have identical frequency response. In addition, the mixers themselves have different conversion loss as well as phase properties. Although the quadrature phase shifter in the LO path of the Q mixer is supposed to be exactly 90°, this is not possible in practice. Furthermore, the power combiner does not perfectly combine the two input signals without providing some imbalance in magnitude and phase. All these imperfections will cause an imbalance between I and Q paths.

8.3 NONIDEAL QUADRATURE MODULATOR

Figure 8.4 illustrates nonideal quadrature modulator, which shows the sources of imperfections inherent in the RF circuitry as well as the DC offsets that can exist at the output of DACs. Nonideal quadrature modulator also includes the low-pass filters to reject aliasing effect of digitally generating the signals. These filters have some frequency-dependent loss and phase shift, which are not necessarily the same for both I and Q channels. In the model, this is shown as $G_{FLI}(\omega)$ and $\phi_{FLI}(\omega)$ for the I channel and $G_{FLQ}(\omega)$ and $\phi_{FLQ}(\omega)$ for the Q channel.

The LO input to the Q mixer is required to be phase shifted by −90°. In most cases, this is accomplished by using a polyphase filter. Another technique that uses a divide-by-two flip-flop achieves better results but requires the LO to be twice the desired frequency [20]. However, in both cases, the phase shift is not exactly −90° and is also frequency dependent. This imperfection is accounted for the model by $\phi_{FLQ}(\omega)$.

Another source of imbalance occurs in the mixing and summing process. I and Q mixers do not have exactly the same conversion loss and phase properties. The RF summing process, which is essentially an RF power combiner, is also not perfectly

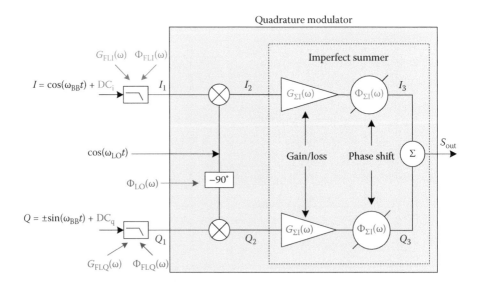

FIGURE 8.4 Nonideal quadrature modulator.

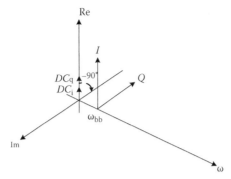

FIGURE 8.5   Vector representation of I and Q signals.

balanced in amplitude and phase. Therefore, there is a certain amount of gain/loss and phase shift, which occurs during the mixing and summing processes, which is different between the I and Q paths. This is represented in our model by assuming the two mixers and the summer to be ideal, and then including the amplitude responses, $G_{\Sigma I}(\omega)$ and $G_{\Sigma Q}(\omega)$, and phase responses, $\phi_{\Sigma I}(\omega)$ and $\phi_{\Sigma Q}(\omega)$, to account for the errors in both processes.

Finally, there is the effect of the DC offsets, which give rise to carrier leakage at the output. These DC offsets are the result of a mismatch of the internal transistors of the quadrature modulator. These DC offsets are included by adding the DC terms DC_i and DC_q to the I and Q channel inputs, respectively.

Using vectors to represent signals is a useful way to visualize the effect of the RF impairments on quadrature modulator performance. Signals can be represented as vectors with a magnitude and phase. Because these vectors will have different frequencies, a third dimension is added to represent frequency. Signals pertaining to the I path and Q paths are illustrated in figures with designations. For instance, the I and Q input signals that are used in the modulator are shown in Figure 8.5.

8.3.1   I PATH

In the nonideal quadrature modulator shown in Figure 8.4, I signal undergoes a phase and amplitude shift through the low-pass filter. This phase shift is, of course, dependent on the baseband frequency of the sinusoid, because the response of the filter is frequency dependent. The DC offset has no phase; so it remains unchanged. The signal can then be expressed as

$$I_1 = G_{FLI}(\omega_{BB}) \cdot \cos(\omega_{BB}t - \varphi_{FLI}(\omega_{BB})) + DC_i \tag{8.5}$$

Signal given in Equation 8.5 can be represented by a vector as shown in Figure 8.6. Mixing I_1 with the LO to obtain the up-converted signal I_2 gives

$$I_2 = \left[G_{FLI}(\omega_{BB}) \cdot \cos(\omega_{BB}t + \varphi_{FLI}(\omega_{BB})) + DC_i \right] \cdot \cos(\omega_{LO}t) \tag{8.6}$$

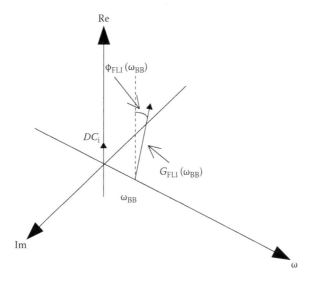

FIGURE 8.6 Vector representation of I_1.

which can be rewritten as

$$I_2 = \frac{G_{FLI}(\omega_{BB})}{2}\left[\cos\left(\omega_U t + \phi_{FLI}(\omega_{BB})\right) + \cos\left(\omega_L t - \phi_{FLI}(\omega_{BB})\right)\right]$$
$$+ DC_i \cdot \cos(\omega_{LO}t) \tag{8.7}$$

where

$$\omega_U = \omega_{BB} + \omega_{LO} \tag{8.8}$$

$$\omega_L = \omega_{BB} - \omega_{LO} \tag{8.9}$$

Vector representation of I_2 is given in Figure 8.7. As described earlier, the summing operation is also imperfect; so it has been represented with some gain and phase shift added. There are three frequency components—each with its own phase shift. Therefore, I_3 can be expressed as

$$I_3 = \frac{G_{FLI}(\omega_{BB})}{2} \cdot \begin{bmatrix} G_{\Sigma I}(\omega_U) \cdot \cos\left(\omega_U t + \phi_{FLI}(\omega_{BB}) + \phi_{\Sigma I}(\omega_U)\right) \\ + G_{\Sigma I}(\omega_L) \cdot \cos\left(\omega_L t - \phi_{FLI}(\omega_{BB}) + \phi_{\Sigma I}(\omega_L)\right) \\ + DC_i \cdot G_{\Sigma I}(\omega_{LO}) \cos\left(\omega_{LO}t + \phi_{\Sigma I}(\omega_{LO})\right) \end{bmatrix} \tag{8.10}$$

I_3 can be set in the following form:

$$I_3 = G_{IU} \cdot \cos\left(\omega_U t + \phi_{IU}\right) + G_{IL} \cdot \cos\left(\omega_L t + \phi_{IL}\right) + G_{LOI} \cdot \cos\left(\omega_{LO}t + \phi_{LOI}\right) \tag{8.11}$$

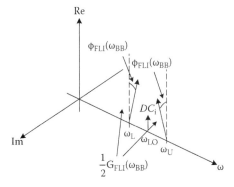

FIGURE 8.7 Vector representation of I_2.

where

$$\phi_{IU} = \phi_{FLI}\left(\omega_{BB}\right) + \phi_{\Sigma I}\left(\omega_{U}\right) \tag{8.12}$$

$$\phi_{IL} = \phi_{\Sigma I}\left(\omega_{L}\right) - \phi_{FLI}\left(\omega_{BB}\right) \tag{8.13}$$

$$G_{IU} = \frac{G_{FLI}\left(\omega_{BB}\right) \cdot G_{\Sigma I}\left(\omega_{U}\right)}{2} \tag{8.14}$$

$$G_{IL} = \frac{G_{FLI}\left(\omega_{BB}\right) \cdot G_{\Sigma I}\left(\omega_{L}\right)}{2} \tag{8.15}$$

$$G_{LOI} = G_{\Sigma I}\left(\omega_{LO}\right) DC_{i} \tag{8.16}$$

and

$$\phi_{LOI} = \phi_{\Sigma I}\left(\omega_{LO}\right) \tag{8.17}$$

The vector representation of I_3 is given in Figure 8.8.

8.3.2 Q PATH

The derivation of the Q signal is identical to the I path signal except that the input signal is now a sine rather than a cosine function. For the sake of derivation, we will consider the case where $Q = +\sin(\omega_{BB}t)$, which would tend to cancel the upper sideband. There is no loss in generality, since we know that by changing the sign of Q, we can cancel the lower sideband instead. As with the I path, the input Q signal undergoes phase and amplitude shifts through the low-pass DAC filter and can be expressed as

$$Q_1 = G_{FLQ}(\omega_{BB}) \cdot \sin\left(\omega_{BB}t - \phi_{FLQ}(\omega_{BB})\right) + DC_q \tag{8.18}$$

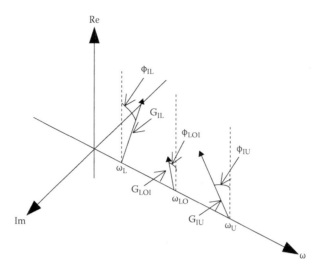

FIGURE 8.8 Vector representation of I_3.

Q_1 can be illustrated in vector form as shown in Figure 8.9. In the LO path, there is a 90° phase shifter as shown in Figure 8.4. However, this phase shift is imperfect, and as a result, an unknown phase error is added, thereby making it a sine wave at the Q mixer with an added phase term ϕ_{LO}. We mix this signal with Q_1 and obtain

$$Q_2 = \left[G_{FLQ}(\omega_{BB}) \cdot \sin\left(\omega_{BB}t - \phi_{FLQ}(\omega_{BB})\right) + DC_q \right] \cdot \sin\left(\omega_{LO}t + \phi_{LO}(\omega_{LO})\right) \quad (8.19)$$

Equation 8.19 can be expressed as

$$Q_2 = \frac{G_{FLQ}(\omega_{BB})}{2} \begin{bmatrix} \cos\left(\omega_L t - \phi_{FLQ}(\omega_{BB}) - \phi_{LO}(\omega_{LO})\right) \\ -\cos\left(\omega_u t - \phi_{FLQ}(\omega_{BB}) + \phi_{LO}(\omega_{LO})\right) \end{bmatrix}$$

$$+ DC_q \cdot \sin(\omega_{LO}t + \phi_{LO}(\omega_{LO})) \quad (8.20)$$

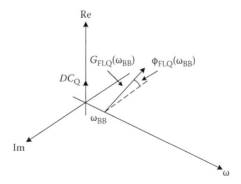

FIGURE 8.9 Vector representation of Q_1.

where ω_U and ω_L are given by Equations 8.8 and 8.9. Q_2 is shown in Figure 8.10 in vector form. When the imperfections in the summation are taken into account as it was done for in phase path, Q_3 can be expressed as

$$Q_3 = \frac{G_{FLQ}(\omega_{BB})}{2} \cdot \begin{bmatrix} G_{\Sigma Q}(\omega_L) \cdot \cos(\omega_L t - \phi_{FLQ}(\omega_{BB}) - \phi_{LO}(\omega_{LO}) + \phi_{\Sigma Q}(\omega_L)) \\ -G_{\Sigma Q}(\omega_U) \cdot \cos(\omega_U t - \phi_{FLQ}(\omega_{BB}) + \phi_{LO}(\omega_{LO}) + \phi_{\Sigma Q}(\omega_U)) \\ +DC_q \cdot G_{\Sigma Q}(\omega_{LO}) \sin(\omega_{LO} t + \phi_{LO}(\omega_{LO})) \end{bmatrix} \quad (8.21)$$

Equation 8.21 can be simplified to

$$Q_3 = G_{QL} \cdot \cos(\omega_L t + \phi_{QL}) - G_{QU} \cdot \cos(\omega_U t + \phi_{QU})$$
$$+ G_{LOQ} \cdot \sin(\omega_{LO} t + \phi_{LOQ}) \quad (8.22)$$

with the following relations:

$$\phi_{QL} = \phi_{\Sigma Q}(\omega_L) - \phi_{FLQ}(\omega_{BB}) - \phi_{LO}(\omega_{LO}) \quad (8.23)$$

$$\phi_{QU} = \phi_{\Sigma Q}(\omega_U) - \phi_{FLQ}(\omega_{BB}) + \phi_{LO}(\omega_{LO}) \quad (8.24)$$

$$G_{QU} = \frac{G_{FLQ}(\omega_{BB}) \cdot G_{\Sigma Q}(\omega_U)}{2} \quad (8.25)$$

$$G_{QL} = \frac{G_{FLQ}(\omega_{BB}) \cdot G_{\Sigma Q}(\omega_L)}{2} \quad (8.26)$$

$$G_{LOQ} = G_{\Sigma Q}(\omega_{LO}) DC_q \quad (8.27)$$

and

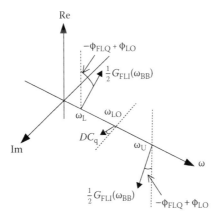

FIGURE 8.10 Vector representation of Q_2.

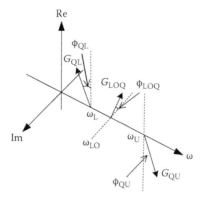

FIGURE 8.11   Vector representation of Q_3.

$$\phi_{LOQ} = \phi_{LO}(\omega_{LO}) \tag{8.28}$$

Q_3 can be depicted as shown in Figure 8.11 in vector form.

8.3.3   OUTPUT OF NONIDEAL QUADRATURE MODULATOR

We now have an expression for the effective I and Q channel signals, including all the imperfections before summation. There are three frequency components on each path that have undergone unknown amplitude and phase shifts and, therefore, significantly degrade the image rejection performance of the system. The summation operation gives S_{out} as

$$I_3 + Q_3 = G_{IU} \cdot \cos(\omega_U t + \phi_{IU}) + G_{IL} \cdot \cos(\omega_L t + \phi_{IL}) + G_{LOI} \cdot \cos(\omega_{LO} t + \phi_{LOI})$$

$$+ G_{QL} \cdot \cos(\omega_L t + \phi_{QL}) - G_{QU} \cdot \cos(\omega_U t + \phi_{QU}) + G_{LOQ} \cdot \sin(\omega_{LO} t + \phi_{LOQ})$$

$$\tag{8.29}$$

There are six sinusoidal terms, but only three frequency components—the upper sideband, the lower sideband, and the LO component. We define output signal as

$$S_{RF} = S_{out} - LO_{out} = RF + IM \tag{8.30}$$

where

$$RF = G_{IL} \cdot \cos(\omega_L t + \phi_{IL}) + G_{QL} \cdot \cos(\omega_L t + \phi_{QL}) \tag{8.31}$$

$$IM = G_{IU} \cdot \cos(\omega_U t + \phi_{IU}) + G_{QU} \cdot \cos(\omega_U t + \phi_{QU}) \tag{8.32}$$

The composite signal S_{out} is the output signal, and it contains both the desired RF component and the undesired image as illustrated in Figure 8.12. In order to get ideal image rejection from a nonideal quad modulator, the phase and amplitude of

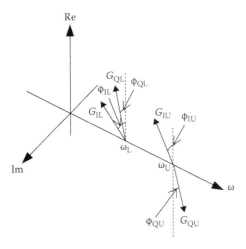

FIGURE 8.12 Vector representation of S_{out}.

the input signals (I and Q) need to be adjusted such that the image vectors at the summed output completely cancel. This requires knowledge of the gain and phase responses of the I and Q channels separately. Furthermore, the degree of accuracy is specifically crucial for the phase.

One of the challenges of improving the image rejection performance is the fact that the I and Q components are inseparable at the output. There have been some innovative calibration methods developed, which measure the power level of the undesired signal as the baseband signals are adjusted to minimize these unwanted components.

Methods such as the ones in Reference. [21] make use of only the magnitude information of the output signal and require an iterative algorithm to arrive at the nominal correction factors, which are impractical.

8.4 IMPROVED METHOD FOR IMAGE BAND REJECTION

In the proposed system [19], a sample of the RF output of the quadrature modulator is mixed with a signal at the image frequency, which is phase-locked with the test signal, thus converting the image component directly to DC. The I and Q paths are then excited separately with a CW tone whose phase is swept from 0 to 2π radians. The resulting DC output is a sinusoidal function with respect to the swept phase of the exciting tone. This sinusoid has amplitude and phase related to the errors imposed in each path. The I and Q test signals are then compared to each other to obtain the correction factors. The simplified block diagram of the proposed system accomplishing the image band rejection is illustrated in Figure 8.13.

Let us consider the case where we wish to null the upper sideband in a quadrature mixer. In this case, we call the upper sideband the image signal and the lower side-band the desired signal. For the first measurement, only I path is excited as shown in Figure 8.14. Because there is no Q signal, there is also no quadrature nulling;

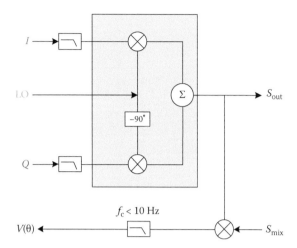

FIGURE 8.13 Block diagram of image band rejection system.

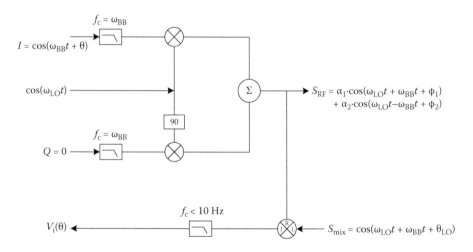

FIGURE 8.14 Image band rejection method with I path test signal measurement.

therefore, both the desired and image signals are present at the output of the quadrature modulator. The output of the system is sampled through a coupler and fed into the RF port of a mixer. The signal at the LO port of the mixer is of the same frequency as the image, thus mixing the image down to DC. A very low-frequency low-pass filter is used at the IF port of the mixer to reject all frequencies except the DC component, which is now a representation of the image signal that we are attempting to null. This can be illustrated using vector signals in Figure 8.15.

When two signals of identical frequencies are applied to a mixer, it produces a DC component at the output, which has a sinusoidal relationship to the phase difference of the two signals as shown in Figure 8.16. This, in essence, corresponds to

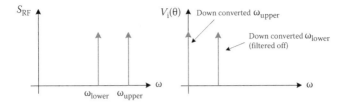

FIGURE 8.15 S_{RF} and $V_i(q)$ signals with I path measurement.

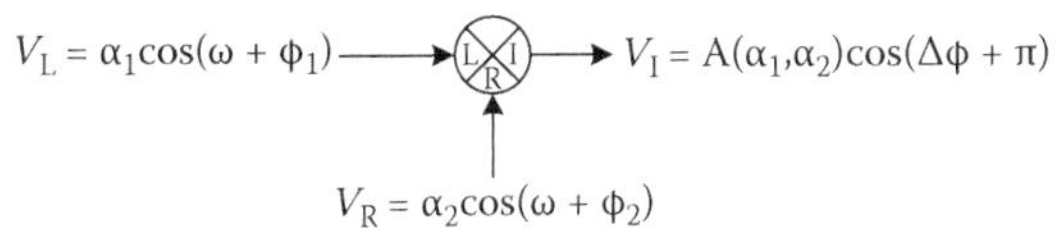

$$V_L = \alpha_1 \cos(\omega + \phi_1) \longrightarrow \boxed{\otimes} \longrightarrow V_1 = A(\alpha_1, \alpha_2)\cos(\Delta\phi + \pi)$$

$$V_R = \alpha_2 \cos(\omega + \phi_2)$$

FIGURE 8.16 Using mixer as phase detector.

using mixer as a phase detector. DC component produced at the output of the mixer depends on magnitudes and relative phase of two sinusoids. In the proposed system, the sum product will be filtered off; so the DC component is the wanted product. Therefore, the resulting DC component seen at IF port is a function of both the phase and magnitude of the driving signals. As the phase of one of signal is swept, the resulting DC values at the IF port will trace out a sinusoid as shown in Figure 8.17.

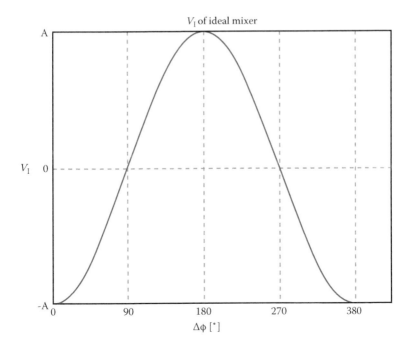

FIGURE 8.17 V_i at the output of a theoretical mixer.

It can be seen from Figure 8.14 that by incrementally adjusting the phase q of the I signal from 0 to 2p and recording DC output, $V_i(q)$, of the mixer, the resulting signal is a cosine function with a phase f_I that is related to the overall change of phase through the I path of the quadrature modulator, and an amplitude a_I is related to the magnitude of the I component at the output.

There is an unavoidable DC offset, which arises from the real mixer itself [35]. This is accounted for by the DC_I term. An extra phase term f_{Offset} is added as well to account for the phase imperfections in the mixer [36]. Because the phase of the LO is unknown, it must also be included. As a result, the resultant signal $V_i(q)$ can be expressed as

$$V_i(\theta) = \alpha_I \cos\left(\theta + \phi_I + \phi_{\text{Offset}} - \theta_{\text{LOi}} + \pi\right) + DC_I \tag{8.33}$$

For the Q path, this process is repeated by keeping I path inactive and exciting only the Q path with a sinusoid in quadrature with the first test signal as shown in Figure 8.18.

The recorded set of DC values $V_q(q)$ is a sinusoid similar to that of the I path; however, it has a different phase f_Q and amplitude a_Q that is related to Q component at the output of the quadrature modulator with DC offset DC_I and phase offset f_{Offset} resulting from imperfections in the mixer. $V_q(q)$ can then be expressed as

$$V_q(\theta) = \alpha_Q \cos\left(\theta + \phi_Q + \phi_{\text{Offset}} - \theta_{\text{LOq}} + \pi\right) + DC_Q \tag{8.34}$$

Because we are using the same LO for both the I and Q signals, the variation in the DC offset will only be affected by the signal present at the RF port. Therefore, DC offsets can easily be removed by subtracting the average from each signal given in Equations 8.33 and 8.34 as

$$V_i(\theta) = \alpha_I \cos\left(\theta + \phi_I + \phi_{\text{Offset}} - \theta_{\text{LOi}} + \pi\right) \tag{8.35}$$

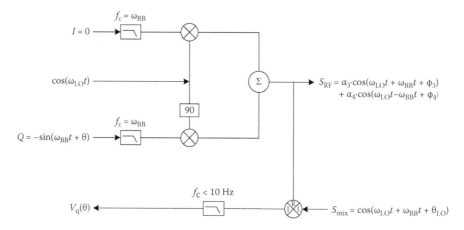

FIGURE 8.18   Image band rejection method with Q path test signal measurement.

$$V_q(\theta) = \alpha_Q \cos\left(\theta + \phi_Q + \phi_{Offset} - \theta_{LOq} + \pi\right) \tag{8.36}$$

As explained earlier, there is also a phase offset induced by the mixer. The effect of this phase offset is that the null at the output of the mixer occurs at some relative phase other than exactly p/2. This phase shift results from the fact that the electrical lengths from L-port to I-port and from R-port to I-port could be different. Sometimes, this is done on purpose to match the ports to 50 Ω. Because the I and Q test signals are identical in frequency, the phase offsets introduced by the mixer will also be identical for both the I and Q measurements. This is because the relative phase and amplitude differences between I and Q paths are of interest, and this allows us to simplify Equations 8.35 and 8.36 to

$$V_i(\theta) = \alpha_I \cos\left(\theta + \phi_I - \theta_{LOi} + \pi\right) \tag{8.37}$$

$$V_q(\theta) = \alpha_Q \cos\left(\theta + \phi_Q - \theta_{LOq} + \pi\right) \tag{8.38}$$

Further simplification of the Equations 8.37 and 8.38 is possible by defining the following equations:

$$\Delta\phi_I = \phi_I - \theta_{LOi} + \pi \tag{8.39}$$

$$\Delta\phi_Q = \phi_Q - \theta_{LOq} + \pi \tag{8.40}$$

So, Equations 8.37 and 8.38 can be expressed as

$$V_i(\theta) = \alpha_I \cos\left(\theta + \Delta\phi_I\right) \tag{8.41}$$

$$V_q(\theta) = \alpha_Q \cos\left(\theta + \Delta\phi_Q\right) \tag{8.42}$$

In order to properly balance the quadrature modulator, the amplitudes and phases of V_i and V_q must be equal. This can be accomplished by only predistorting one of the channels. When Q channel is predistorted, we can define the amplitude and phase correction factors for the Q path as

$$\alpha_{corr} = \frac{\alpha_I}{\alpha_Q} \tag{8.43}$$

$$\phi_{corr} = \Delta\phi_I - \Delta\phi_Q \tag{8.44}$$

The correction factors are then applied to the transmission signal as follows:

$$I = \cos\left(\omega_{BB}t\right) \tag{8.45}$$

$$Q = \alpha_{corr} \sin\left(\omega_{BB}t - \phi_{corr}\right) \tag{8.46}$$

8.4.1 LO CANCELLATION

In this section, the method for canceling the LO leakage in a quadrature modulator is proposed. For the clarity of the analytical derivation, we consider only the LO components in the system. As described previously, DC offsets are added to the I and Q paths to cancel the offsets that exist to create the LO leakage. The goal of the system is to find the correct DC offset for each channel. The LO cancellation system shown in Figure 8.19 is similar to the one proposed for the image cancellation system by down-converting the undesired signal to DC. By incrementally changing the phase on S_{mix} from 0 to 2π radians and monitoring the resulting DC output at V_{LO}, a relative measure of the phase and magnitude of the LO signal can be obtained. The process of using a mixer as a phase detector has already been discussed in the previous section in detail. The offsets that are desired to be cancelled are represented by DC_i and DC_q, and the factors ΔI and ΔQ are added to accomplish this. The goal is to determine the values of these correction factors.

The system in Figure 8.19 can be mathematically modeled in a slightly different manner. Because all of the signals in the system are phase-locked, it is assumed that the time-varying aspect of the sinusoids in the system can be ignored. From the analysis conducted in Section 8.3 for an unbalanced quadrature modulator, the LO components at the output were found to be

$$\text{LO}_{out} = G_{LOQ} \cdot \sin(\omega_{LO}t + \phi_E) + G_{LOI} \cdot \cos(\omega_{LO}t + \phi_{LOI}) \qquad (8.47)$$

The expression can be simplified by defining

$$G_{LOI} = G_{\Sigma I}(\omega_{LO})DC_i, \qquad (8.48a)$$

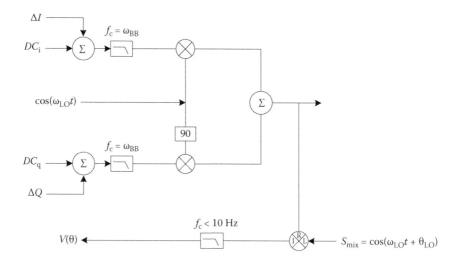

FIGURE 8.19 Quadrature modulator LO cancellation system.

and

$$G_{\text{LOQ}} = G_{\Sigma Q}(\omega_{\text{LO}})DC_q. \qquad (8.48b)$$

Then, Equation 8.47 can be expressed as

$$\text{LO}_{\text{out}} = G_{\Sigma Q}(\omega_{\text{LO}})DC_q \cdot \sin(\omega_{\text{LO}}t + \phi_{\text{LOQ}}) + G_{\Sigma I}(\omega_{\text{LO}})DC_i \cdot \cos(\omega_{\text{LO}}t + \phi_{\text{LOI}}) \qquad (8.49)$$

In Figure 8.19, the terms denoted by ΔI and ΔQ are the correction factors added to cancel DC_i and DC_q. Then, Equation 8.49 can be modified to include these correction factors as

$$\text{LO}_{\text{out}} = (DC_i + \Delta I)G_{\Sigma I}(\omega_{\text{LO}}) \cdot \cos(\omega_{\text{LO}}t + \phi_{\text{LOI}})$$
$$+ (DC_q + \Delta Q)G_{\Sigma Q}(\omega_{\text{LO}}) \cdot \sin(\omega_{\text{LO}}t + \phi_{\text{LOQ}}) \qquad (8.50)$$

Equation 8.50 can be put into vector representation as

$$\begin{bmatrix} \text{LO}_x \\ \text{LO}_y \end{bmatrix} = [R_{\text{I}}]\begin{bmatrix} 1 \\ 0 \end{bmatrix}DC_i + [R_{\text{I}}]\begin{bmatrix} 1 \\ 0 \end{bmatrix}\Delta I + [R_{\text{Q}}]\begin{bmatrix} 0 \\ -1 \end{bmatrix}DC_q + [R_{\text{Q}}]\begin{bmatrix} 0 \\ -1 \end{bmatrix}\Delta Q$$

$$(8.51)$$

where

$$[R_{\text{I}}] = \begin{bmatrix} G_{\Sigma I}(\omega_{\text{LO}}) \cdot \cos(\phi_{\text{LOI}}) & -G_{\Sigma I}(\omega_{\text{LO}}) \cdot \sin(\phi_{\text{LOI}}) \\ G_{\Sigma I}(\omega_{\text{LO}}) \cdot \sin(\phi_{\text{LOI}}) & G_{\Sigma I}(\omega_{\text{LO}}) \cdot \cos(\phi_{\text{LOI}}) \end{bmatrix} \qquad (8.52)$$

$$[R_{\text{Q}}] = \begin{bmatrix} G_{\Sigma Q}(\omega_{\text{LO}}) \cdot \cos(\phi_{\text{LOQ}}) & -G_{\Sigma Q}(\omega_{\text{LO}}) \cdot \sin(\phi_{\text{LOQ}}) \\ G_{\Sigma Q}(\omega_{\text{LO}}) \cdot \sin(\phi_{\text{LOQ}}) & G_{\Sigma Q}(\omega_{\text{LO}}) \cdot \cos(\phi_{\text{LOQ}}) \end{bmatrix} \qquad (8.53)$$

The expression for LO component at the output of the quadrature modulator given in Equation 8.51 includes all of the RF impairments and offset corrections applied to the circuit.

As described in the image cancellation process, the output signal is down-converted directly to DC. This process allows us to low-pass filter all other components of the output signal and observe only the LO leakage. By sweeping the phase of S_{mix} from 0 to 2π and measuring the resulting DC output of the mixer with respect to the phase of S_{mix}, we are using the mixer as a phase and magnitude detector as illustrated in Figure 8.19.

There is some relative phase and amplitude error in down-conversion process when physical devices are used. However, this error can also be modeled as a simple rotation matrix $[M]$ as shown in Figure 8.20. We can now define our signal V as

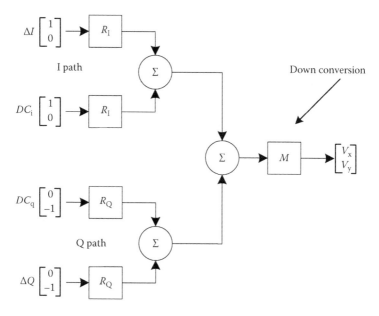

FIGURE 8.20 Linear algebraic model for LO cancellation system.

$$\left[\begin{array}{c} V_x \\ V_y \end{array}\right]=[M]\left([R_I]\left[\begin{array}{c} 1 \\ 0 \end{array}\right]DC_i+[R_I]\left[\begin{array}{c} 1 \\ 0 \end{array}\right]\Delta I+[R_Q]\left[\begin{array}{c} 0 \\ -1 \end{array}\right]DC_q+[R_Q]\left[\begin{array}{c} 0 \\ -1 \end{array}\right]\Delta Q\right)$$

(8.54)

We define the following terms to simplify Equation 8.54 as

$$[M][R_I]=[A]$$

(8.55a)

and

$$[M][R_Q]=[B]$$

(8.55b)

Then, Equation 8.54 can be expressed as

$$\left[\begin{array}{c} V_x \\ V_y \end{array}\right]=[A]\left[\begin{array}{c} 1 \\ 0 \end{array}\right](DC_i+\Delta I)+[B]\left[\begin{array}{c} 0 \\ -1 \end{array}\right](DC_q+\Delta Q)$$

(8.56)

The simplified Equation 8.56 can be modeled as shown in Figure 8.21.

In the system shown in Figure 8.21, we have control only over ΔI and ΔQ, because these are the values we are using to null the LO leakage. In addition, because we are measuring the output of our system, V_x and V_y are known as well. However, the matrices $[A]$ and $[B]$, or the scalars DC_i and DC_q are unknowns. The unknown

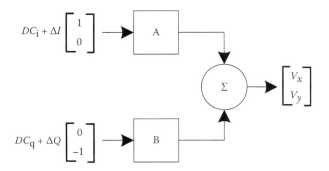

FIGURE 8.21   Simplified mathematical model of LO cancellation system.

terms can be determined using the rotational matrices $[A]$ and $[B]$ and solving the relation given in Equation 8.56, which is set in the following form:

$$\begin{bmatrix} V_x \\ V_y \end{bmatrix} = \begin{bmatrix} a_{11} & -a_{12} \\ a_{12} & a_{11} \end{bmatrix} \begin{bmatrix} 1 \\ 0 \end{bmatrix} (DC_i + \Delta I) + \begin{bmatrix} b_{11} & -b_{12} \\ b_{12} & b_{11} \end{bmatrix} \begin{bmatrix} 0 \\ -1 \end{bmatrix} (DC_q + \Delta Q)$$

(8.57)

The Equation 8.57 is a system of two equations with six unknowns, where a minimum of three measurements is needed to solve for all of the unknowns. This is shown by

$$V_x = a_{11}DC_i + a_{11}\Delta I + b_{12}DC_q + b_{12}\Delta Q \tag{8.58}$$

$$V_y = a_{12}DC_i + a_{12}\Delta I - b_{11}DC - b_{11}\Delta Q \tag{8.59}$$

The unknown terms are derived by solving Equations 8.58 and 8.59 and are obtained as

$$DC_i = \frac{b_{12}K_y + b_{11}K_x}{a_{11}b_{11} + a_{12}b_{12}} \tag{8.60}$$

$$DC_q = \frac{a_{12}K_x - a_{11}K_y}{a_{11}b_{11} + a_{12}b_{12}} \tag{8.61}$$

where

$$a_{11} = \frac{V_{x1} - V_{x2}}{\Delta I_1 - \Delta I_2} \tag{8.62a}$$

$$a_{12} = \frac{V_{y1} - V_{y2}}{\Delta I_1 - \Delta I_2} \tag{8.62b}$$

$$b_{11} = \frac{V_{y1} - V_{y3}}{\Delta Q_2 - \Delta Q_1} \qquad (8.62c)$$

$$b_{12} = \frac{V_{x1} - V_{x3}}{\Delta Q_1 - \Delta Q_2} \qquad (8.62d)$$

and

$$K_x = V_x - a_{11}\Delta I - b_{12}\Delta Q \qquad (8.63a)$$

$$K_y = V_y - a_{12}\Delta I + b_{11}\Delta Q \qquad (8.63b)$$

LO leakage can now be cancelled by setting $\Delta I = -DC_i$ and $\Delta Q = -DC_q$ in LO cancellation system shown in Figure 8.19.

8.4.2 SIMULATION RESULTS

Simulation is performed using Agilent's ADS software with harmonic balance simulation to validate the proposed method. A model for the quadrature modulator, including the RF imperfections discussed, is created in ADS as shown in Figure 8.22.

8.4.2.1 Simulation Model

Phase and gain imbalances were added to the I and Q channels in order to show degraded image rejection. DC offsets were also added into I and Q paths, which create the LO leakage. The values of these imbalances are typical for quadrature modulators in the market today.

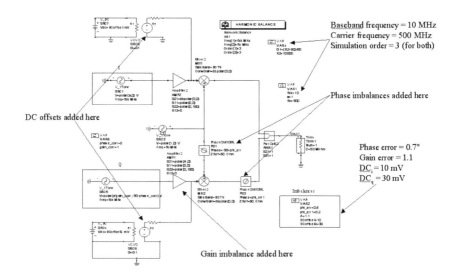

FIGURE 8.22 Simulation model of an unbalanced quadrature modulator in ADS.

The input signal for the I channel is a CW tone at 10 MHz with a phase of 0° (cosine), and the Q input signal is a CW tone at the same frequency, which is shifted by −90° (sine). The LO source has a frequency of 500 MHz with 0° phase. Because the Q signal is a positive sine function, the quadrature modulator is configured to cancel the upper sideband (510 MHz). The desired RF signal is the lower sideband (490 MHz). For simplicity, all imbalances were added to the Q side. Adding imbalances to only one side is valid since the image rejection depends only on the relative imbalance between the two channels. The nominal 90° LO phase shifter for the Q channel has 0.5° of phase error. There is also another 0.2° of phase error added to the Q side before RF combiner to simulate the phase error in the summing circuit. An amplitude imbalance of 1.1 is also added to the Q side as variable A. For the LO leakage, 10 mV of DC offset is added to the I channel and 30 mV of DC offset is added to the Q channel. The imbalances and DC offsets added into this simulation are typical of most commercial quadrature modulators available for designers in the market as mentioned earlier. For this reason, measurements are normally taken to improve this performance by predistorting the input signals to account for these imbalances and offsets. This can be accomplished by a calibration process where the spectrum is monitored as the phases, amplitudes, and DC offsets are manipulated to obtain the best result.

This lengthy process is performed at many frequencies and temperatures depending on the application. The ability to null the image and LO is dependent on the process of using a mixer as a phase detector. This process was modeled and simulated in ADS as illustrated in Figure 8.23.

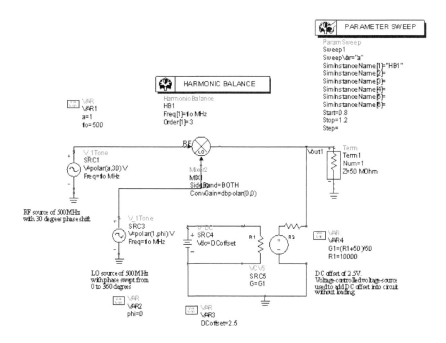

FIGURE 8.23 Simulation model of mixer as phase detector using ADS.

Two identical frequencies are applied to the LO and RF ports of a mixer in Figure 8.23. The signal at the RF port has a phase shift of 30°. The phase of the LO signal is swept from 0° to 360° in 1° increments. The amplitude of the RF signal is also swept from 0.8 to 1.2 V in 0.1 V increments. Because the mixer model in ADS does not include the DC offset that a physical mixer would have, an external DC offset of 2.5 V was applied to the output of the mixer through a voltage-controlled voltage source. When the phase of one of the input signals is swept and the DC output of the mixer versus phase is charted, the relative phase of the other input signal can be determined.

In addition, the magnitude of the resulting sinusoid is a function of the magnitude of the RF input signal. Furthermore, the inherent DC offset can be determined by simply taking the average value across all phases.

The simulation of the image cancellation system is a version of the unbalanced quadrature modulator simulation shown in Figure 8.22. There are two identical circuits—one of them simulates I test signal and the other simulates the Q test signal. The imbalances added into both circuits are identical to the unbalanced quadrature modulator simulation. The simulator sweeps the phases of both input signals from 0° to 360° in 0.1° increments. In both cases, the DC output of the mixer is captured in the same manner as described in the Section 8.4.2.1 where the mixer as a phase

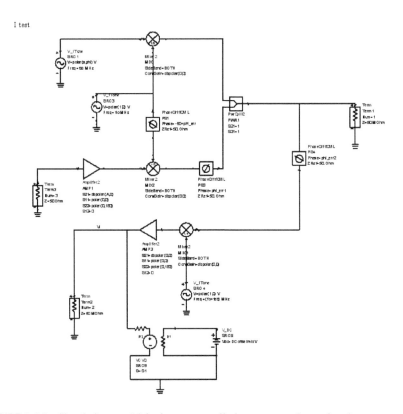

FIGURE 8.24 Simulation model for image cancellation system—I test signal.

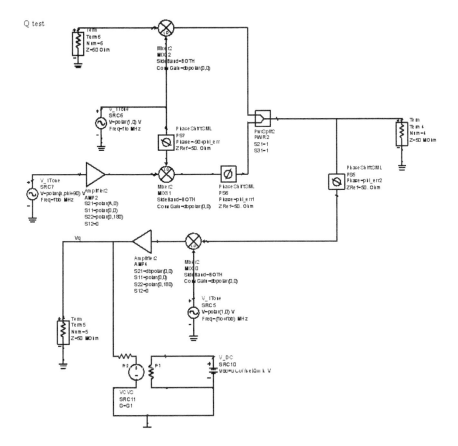

FIGURE 8.25   Simulation model for image cancellation system—Q test signal.

detector was simulated. In addition, different DC offsets were applied to each of the output mixers in order to simulate the DC offset of a real mixer. The simulation models for image cancellation using I test signal and Q test signal are illustrated in Figures 8.24 and 8.25, respectively.

8.4.2.2   Simulation Results

The simulation results of an unbalanced quadrature modulator shown in Figure 8.22 have the significant LO leakage and image signal present at the output without the implementation of the proposed method as illustrated Figure 8.26. The simulation setup used by ADS is illustrated in Table 8.1.

The harmonic balance simulator produces data for all product frequencies specified by the user in the simulation setup. In our case, there are two identical frequencies mixing together to produce DC, twice the frequency, three times the frequency, and so on. Since it is assumed that a low-pass filter will be used to strip off all components except for DC, the simulation results show only the DC component.

As can be seen from Figure 8.27, the DC output of the mixer, which is functioning as a phase detector using the model shown in Figure 8.23, is a cosine function

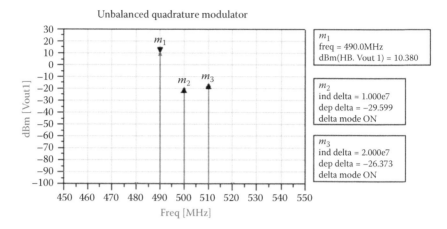

FIGURE 8.26 Unbalanced quadrature modulator simulation results.

TABLE 8.1
Simulation Setup for Image Cancellation System

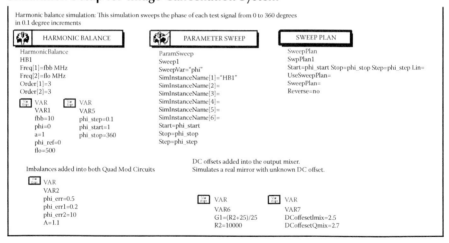

with respect to the swept phase variable of the test signal. Each trace corresponds to a separate input power of the test signal. The simulation results show that the output magnitude of the resulting sinusoid is a function of the test signal input power.

By taking the maximum point on each trace shown in Figure 8.27, the phase of the output signal is determined. This cosine has a phase of $210°$, and since we know from Figure 8.16 that the relative phase of the two signals can be determined by subtracting π radians from this cosine function, it has been successfully determined that the input signal's phase is $30°$. To prove that the DC offset can be removed, an average of each trace was taken and compared with the DC offset of $2.5\,V$ applied in the simulation. In each case, the DC offset was determined to be $2.5\,V$.

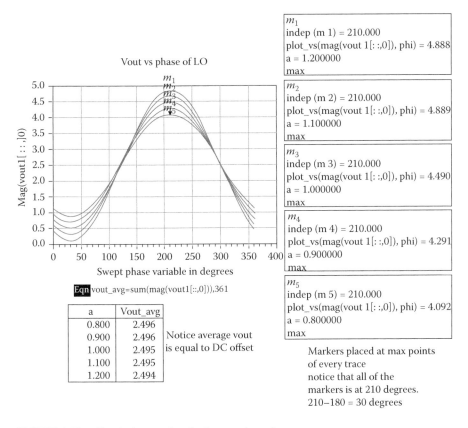

Vout vs phase of LO

a	Vout_avg
0.800	2.496
0.900	2.496
1.000	2.495
1.100	2.495
1.200	2.494

Eqn vout_avg=sum(mag(vout1[::,0])),361

Notice average vout is equal to DC offset

m_1
indep (m 1) = 210.000
plot_vs(mag(vout 1[: :,0]), phi) = 4.888
a = 1.200000
max

m_2
indep (m 2) = 210.000
plot_vs(mag(vout 1[: :,0]), phi) = 4.889
a = 1.100000
max

m_3
indep (m 3) = 210.000
plot_vs(mag(vout 1[: :,0]), phi) = 4.490
a = 1.000000
max

m_4
indep (m 4) = 210.000
plot_vs(mag(vout 1[: :,0]), phi) = 4.291
a = 0.900000
max

m_5
indep (m 5) = 210.000
plot_vs(mag(vout 1[: :,0]), phi) = 4.092
a = 0.800000
max

Markers placed at max points
of every trace
notice that all of the
markers is at 210 degrees.
210–180 = 30 degrees

FIGURE 8.27 Simulation results of mixer as phase detector.

The DC output of the mixers are captured as V_i and V_q for the I and Q test simulations, respectively, as shown in Figure 8.28. The phase of each test signal was found by simply finding the value of the phase variable at the maximum of each trace. The magnitudes of the test signals were found by the following formula:

$$mag_{i,q} = \frac{\left(\max\{V_{i,q}\} - \min\{V_{i,q}\}\right)}{2} \tag{8.64}$$

For the first set of data, the DC offset was left in the calculations. Gain and phase correction factors were found by Equations 8.43 and 8.44.

For the second set of data, the DC offset was removed by taking the average value of both traces. All other values were calculated in the exact same manner as described earlier. The simulation results for this case are shown in Figure 8.29.

In both cases, DC offset included and DC offset removed, identical correction factors were obtained. The obtained values agree with the amount of error added to the simulation; for instance, the correction was 0.7°. This agrees with the 0.5° added to the LO phase shifter plus the 0.2° added to the summing block. The gain

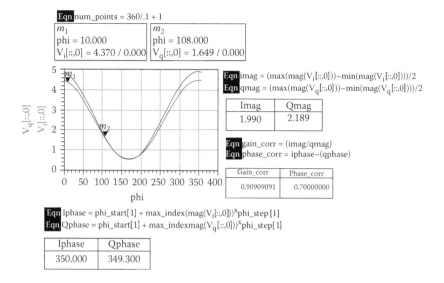

FIGURE 8.28   Simulation results for quadrature modulator image cancellation simulation results—DC offset included.

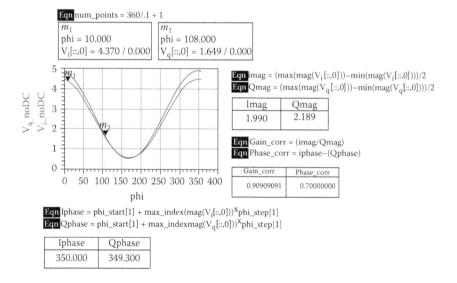

FIGURE 8.29   Simulation results for quadrature modulator image cancellation simulation results—DC offset included.

correction was calculated to be 0.90909, which is exactly the inverse of the amplitude error of 1.1 that was added.

After correction factors are obtained, the validity of the proposed system is checked again by resimulating the unbalanced quadrature modulator with the correction factors applied. The simulation was performed, and the correction factors were applied in accordance with Equations 8.45 and 8.46. The simulation model with the correction factors applied for image cancellation is shown in Figure 8.30. Please note that the lower left corner in Figure 8.30 shows the application of the correction factors. Simulation results show that by predistorting the Q channel with the appropriate correction factors based on the proposed image cancellation method, the system shows an improvement from 26 to 226 dB of image rejection as illustrated in Figure 8.31 and confirms the analytical results derived.

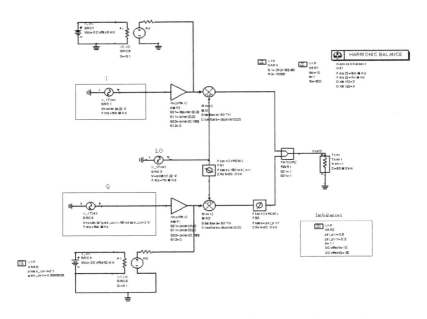

FIGURE 8.30 Simulation model for image cancellation system with correction.

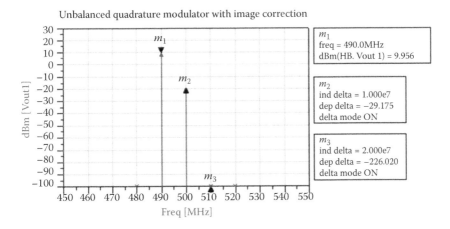

FIGURE 8.31 Simulation results for an unbalanced quadrature modulator with correction factors.

The application of correction factors, which are obtained from two sets of measurements, drastically reduces the unwanted image signal level designated by m3 as shown in Figure 8.31.

The possible hardware implementation of the proposed system is illustrated in Figure 8.32. In the proposed implementation, I and Q signals are created digitally in an FPGA and converted to analog signals through the D/A converters. Low-pass filters are necessary to mitigate aliasing. All analog signal sources need to be phase-locked. This can be done using PLLs with a common reference. LO source, which drives the quad modulator, will define the output phase noise and tune time of the overall system. The other signal sources are only needed for down-conversion. There are broadband PLL ICs available on the market today, which have integrated VCOs in a single small package with very low power consumption. A/D converters are responsible for converting the DC output from the sample mixers to a digital signal. Because these devices are only processing a DC signal (or low frequency), there is no need to use a broadband communications A/D. These data converters would most likely be small and consume little power. FPGA is responsible for creating the digital I and Q signals and collecting the samples LO and image signals. All calculations for the appropriate correction factors and DC offsets are performed inside the FPGA. This process is simply an algebraic operation, and it does not involve any lengthy algorithm.

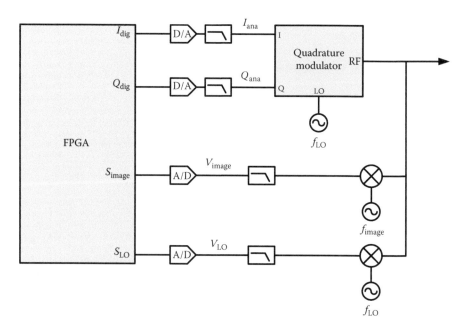

FIGURE 8.32 Proposed hardware implementation of quad modulator balancing system.

REFERENCES

1. B. Razavi, *RF Microelectronics*, Prentice Hall, Upper Saddle River, NJ, 1998.
2. IEEE Std 802.11 a-1999, Part 11: Wireless LAN medium access control (MAC) and physical layer (PHY) specifications: High-speed physical layer in the 5-GHz band, December 1999.
3. Y.-J. Ko and S. Stapleton, Gain and phase mismatch effects on double image rejection transmitter, *IET Circuits, Devices and Systems*, Vol. 5, No. 3, pp. 212–221, 2011.
4. A. Georgiadis, Gain, phase imbalance, and phase noise effects on error vector magnitude, *IEEE Transactions on Vehicular Technology*, Vol. 53, No. 2, pp. 443–449, 2004.
5. A. Baier, Quadrature mixer imbalances in digital TDMA mobile radio receivers, in *Proceedings of International Zurich Seminar on Digital Communications, Electronic Circuits and Systems for Communications*, pp. 147–162, March 1990.
6. C. L. Liu, Impacts of I/Q imbalance on QPSK-OFDM-QAM detection, *IEEE Transactions on Consumer Electronics*, Vol. 44, pp. 984–989, 1998.
7. B. Come, R. Ness, S. Donnay, L. Van der Perre, W. Eberle, P. Wambacq, M. Engels, and I. Bolsens, Impact of front-end non-idealities on bit error rate performance of WLAN-OFDM transceivers, in *Proceedings of IEEE Radio and Wireless Conference (RAWCON)*, pp. 91–94, 2000.
8. J. Mahattanakul, The effect of I/Q imbalance and complex filter component mismatch in low-IF receivers, *IEEE Transactions on Circuits and Systems I, Regular Papers*, Vol. 53, No. 2, pp. 247–253, 2006.
9. A. Tarighat, R. Bagheri, and A. H. Sayed, Compensation schemes and performance analysis of IQ imbalances in OFDM receivers, *IEEE Transactions on Signal Processing*, Vol. 53, No. 8, pp. 3257–3268, 2005.
10. J. Tubbax, B. Come, L. V. der Perre, S. Donnay, M. Engels, M. Moonen, and H. D. Man, Joint compensation of IQ imbalance and frequency offset in OFDM systems, in *Proceedings of Radio and Wireless Conference*, pp. 39–42, August 2003.
11. A. Schuchert, R. Hasholzner, and P. Antoine, A novel IQ imbalance compensation scheme for the reception of OFDM signals, *IEEE Transactions on Consumer Electronics*, Vol. 47, pp. 313–318, 2001.
12. S. Fouladifard and H. Shafiee, Frequency offset estimation in OFDM systems in presence of IQ imbalance, in *Proceedings of International Conference on Communications*, pp. 2071–2075, May 2003.
13. A. Tarighat and A. H. Sayed, MIMO OFDM receivers for systems with IQ imbalances, *IEEE Transactions on Signal Processing*, Vol. 53, No. 9, pp. 3583–3596, 2005.
14. J. Lin and E. Tsui, Joint adaptive transmitter/receiver IQ imbalance correction for OFDM systems, *IEEE International Symposium on PIMRC*, Vol. 2, pp. 1511–1516, September 5–8, 2004.
15. K.-H. Lin, H.-L. Lin, S.-M. Wang, and R. C. Chang, Implementation of digital IQ imbalance compensation in OFDM WLAN receivers, *2006 IEEE International Symposium on Circuits and Systems*, November 2006.
16. L. Moult and J. E. Chen, The K-model: RFIC modeling for communication systems simulation, *IEE Colloquium on Analog Signal Processing (Ref. No. 1998/472)*, pp. 11/1–11/8, October 28, 1998.
17. W. V. Moer and Y. Rolain, Determining the reciprocity of mixers through 3-port large signal network analyzer measurement, *62nd ARFTG Conference Digest*, pp. 165–170, December 4–5, 2003.
18. S. K. Myoung, X. Cui, P. Roblin, D. Chaillot, F. Verbeyst, M. Vanden Bossche, S. J. Doo, and W. Dai, Large signal network analyzer with trigger for baseband modulation linearization, *64th ARFTG Conference Digest*, Orlando, FL, December 2004.

19. A. Eroglu, Non-invasive quadrature modulator balancing method to optimize image band rejection, *IEEE Transactions on Circuits and Systems I, Regular Papers*, Vol. 61, No.2, pp. 600–612, 2014.

20. Y.-H. Hsieh, W.-Y. Hu , S.-M. Lin, C.-L. Chen, W.-K. Li, S.-J. Chen, and D. J. Chen, An auto-I/Q calibrated CMOS transceiver for 802.11g, *IEEE Journal of Solid-State Circuits*, Vol. 40, No. 11, pp. 2187–2192, 2005.

21. Analog Devices, Inc., ADL-5385, Datasheet, 2006. Norwood, MA.

22. J. Craninckx, B. Debaillie, B. Come and S. Donnay "A WLAN direct up-conversion mixer with automatic image rejection calibration," *Int. Solid-State Circuits Conf. Tech. Dig.*, pp. 546–616, 2005.

23. I. Vassiliou "A single-chip digitally calibrated 5.15-5.825-GHz 0.18-mm CMOS transceiver for 802.11a wireless LAN," *IEEE J. Solid-State Circuits*, vol. 38, no. 12, pp. 2221–2231, 2003.

24. C.P., Lee, A., Behzad, D., Ojo, M., Kappes, S., Au, P., Meng-An K., Carter,S., Tian, "A Highly Linear Difrect-Conversion Transmit Mixer Transconductance Stage with Local Oscillation Feedthrough and I/Q Imbalance Cancellation Scheme," *Int. Solid-State Circuits Conf. Tech. Dig.* , pp. 1450–1459, 2006.

25. L. Ding, Z. Ma, D. R. Morgan, M. Zierdt, and G. T. Zhou, "Compensation of Frequency-Dependent Gain/Phase Imbalance in Predistortion Linearization Systems," *IEEE Trans. Circuits and Syst. I, Reg. Papers*, vol. 55, no.1, pp. 390–397, Feb. 2008.

26. L. Yu, and W.M. Snelgrove "A novel adaptive mismatch cancellation system for quadrature IF radio receivers," *IEEE Trans. Circuits and Syst. II, Analog Digit. Signal Process*, vol. 46, no.6, pp. 789–801, June 1999.

27. Olli Myllari, Lauri Anttila, Mikko Valkama, "Digital transmitter I/Q imbalance calibration: real-time prototype implementation and performance measurement," 18th Proc of EUSIPCO, pp. 537–541, August 23–27, 2010.

28. R. Montemayor and B. Razavi, "A self-calibrating 900-MHz CMOS image-reject receiver," Proceedings of the 26rd ESSCIRC, pp. 320–323, 19–21 Sept. 2000.

Index

Page numbers followed by *f* indicate figures; those followed by *t* indicate tables.

T - #0996 - 101024 - C376 - 234/156/20 - PB - 9781138745773 - Gloss Lamination